Springer Proceedings in Archaeology and Heritage

This conference proceedings series publishes the latest research developments in all areas of archaeology and heritage (both natural and cultural, tangible and intangible).

This series welcomes an international and diverse range of submissions from conferences, workshops and similar scientific meetings. It aims at disseminating high quality scientific information with an emphasis on cutting-edge findings and will constitute a comprehensive source of reference on a field or subfield of relevance in archaeology and heritage.

Proposals must include the following:

- Name, place and date of the scientific meeting
- A link to the committees (local organization, international advisors etc.)
- Scientific description of the meeting
- List of invited/plenary speakers
- A table of contents including authors and the titles of their respective papers
- A description of the review process for the accepted papers that were presented at the conference
- An estimate of the planned proceedings book parameters (number of pages/ articles, requested number of bulk copies (if applicable), interest in gold open access publication, submission deadline))

For information on submitting a proposal, please contact the Springer responsible editor shown on the series webpage (see "Contacts").

Ángel F. Perles-Ivars · Laura Fuster-López ·
Emanuela Bosco
Editors

Collection Care

Environmental Monitoring, Risk Assessment and Risk Management

 Springer

Editors
Ángel F. Perles-Ivars
Instituto ITACA
Universitat Politècnica de València
Valencia, Spain

Laura Fuster-López
Instituto Universitario de Restauración del
Patrimonio
Universitat Politècnica de València
Valencia, Spain

Emanuela Bosco
Department of the Built Environment
Eindhoven University of Technology
Eindhoven, Noord-Brabant
The Netherlands

ISSN 3059-300X   ISSN 3059-3018   (electronic)
Springer Proceedings in Archaeology and Heritage
ISBN 978-3-031-85654-9   ISBN 978-3-031-85655-6   (eBook)
https://doi.org/10.1007/978-3-031-85655-6

This Springer imprint is published by the registered company Springer Nature Switzerland AG
The registered company address is: Gewerbestrasse 11, 6330 Cham, Switzerland

If disposing of this product, please recycle the paper.

Preface

Over the last decade, the discipline of preventive conservation has experienced a significant step forward due to synergistic technical, technological and conceptual advances. On the one hand, the development of innovative sensor technologies and monitoring systems has made it possible to access accurate, real-time information on the condition of cultural objects during exhibition, storage and/or transport. On the other hand, the study of material properties and their response to environmental agents has led to the development of a significant number of predictive models aimed at estimating potential risks in collections and anticipating corrective measures to mitigate (or even prevent) damage. Finally, the development of user-friendly platforms and analysis tools has led to wider access to information for GLAM (galleries, libraries, archives and museums) stakeholders. Such tools help establish solid communication channels between scientists and the conservation discipline, which are essential for implementing informed (and successful) conservation strategies.

This book aims to provide a comprehensive overview of this developing and innovative research field. To this end, 21 contributions have been selected. The book is divided into two sections: the first brings together 13 papers that focus on the monitoring, analysis, data interpretation and modelling of the effects of different degradation agents (temperature, relative humidity, pollutants, light, vibration, etc.) in materials present in cultural heritage objects. The second part of the book presents eight case studies dealing with risk assessment and risk management in different types of collections with diverse needs, priorities and resources, thus illustrating the complexity of implementing all the current knowledge on sensing, monitoring, predictive analysis and preventive conservation in the design of conservation strategies. The case studies presented also demonstrate how these tools have contributed to maximise dialogue and coordination between the different stakeholders involved in the conservation of cultural heritage.

The first chapter, by **Beltran et al.**, presents the degradation-focused curriculum developed as part of the *Managing Collection Environments Initiative* at the Getty Conservation Institute, aimed to promote the analysis of temperature and relative humidity data, firstly, establishing a baseline of knowledge on the fundamentals of data analysis and visualisation, and secondly reviewing of existing online temperature and relative humidity analysis tools, their application and interpretation. **King et al.** explain the potential of *Conserv* software, a technology company focused on preventive conservation tools, which allows observations to be linked alongside time series visualisations, provides users with the ability to enter both quantitative and qualitative data. Additionally,and the software assigns categories for qualitative observations, based on the idea that the more data collected on the direct effects of the environment on collections over time, the better the understanding of the environmental effects on materials through artificial ageing techniques. In collaboration with colleagues in the Operations Department and the building automation company TEC Systems, Inc. **Haddad et al.** developed a web-based platform to monitor, evaluate, receive alerts and report on environmental conditions at the *Museum of Modern Art* (MOMA) (e.g. gallery and non-gallery spaces, on-site storage, archives and libraries, conservation studios, workshops and offices, etc.), which has facilitated and enriched dialogue between conservation and operations teams, provided tools to better understand and respond to the museum environment, and raised awareness of collection care needs and requirements. The paper by **Kramer et al.** discusses the extent to which the evolution of standards and guidelines has changed building climate control systems and presents the energy savings achieved by dynamic climate control (a concept that refers to controlled variations to achieve an optimal course of temperature and relative humidity throughout the year, considering collection and comfort requirements) compared to static climate control. **Shah and Long** present how data science tools fit into the V&A's collections management strategy and present some methods for transforming, modelling and translating environmental data into insights that can be communicated to museum stakeholders.

In addition to papers focusing on temperature and relative humidity, some chapters discuss light, pollutants and vibration. For example, **Zarzo et al.** present the advantages and disadvantages of zenithal lighting used in museums and other heritage sites, evaluating the transmittance of visible, ultraviolet and near infrared solar radiation through the large skylight that illuminates much of the interior of the l'Almoina Archaeological Museum (Valencia, Spain). The sustainability of lamps and lighting systems in terms of reducing, reusing and recycling materials is discussed by **Saunders**, who proposes a model for assessing the impact of the various changes in museum lighting practice introduced to improve environmental performance on different measures of sustainability, in particular their impact on social and societal sustainability, which includes the longer-term conservation of objects. **Alcayde and Tapol** review the research carried out in recent decades on indoor pollutants in museums, internal emissions, material selection and mitigation to develop preventive conservation strategies. Finally, **Alami et al.** present the data acquisition system developed to record both sound and vibrations simultaneously, to assess the potential for sound-induced vibrations to be transmitted to heritage objects inside the display case.

The following four chapters on analysis and experimental modelling conclude this first group of papers and act as a bridge between the two sections, linking research and technology to conservation practice. **Thickett** explains the challenges posed by the continuous measurement of RH fluctuations, where correlation with observed physical changes is limited and where techniques such as acoustic emission have proved very effective. **Sheets et al.** show that advanced analytical approaches using a combination of engineering and convolutional neural network techniques can be effective in identifying and quantifying the effects of surface ageing in Asian lacquers, thereby improving the establishment of appropriate conservation strategies. **Stöcklein et al.** present a tool for damage and risk assessment of conservation and restoration treatments on wooden art objects and investigate the interaction between wood and varnish layers, developed using the finite element simulation method in a hygromechanical model as part of the Cultwood research project. **Jeberien et al.** present the possibilities and limitations of the Oddy test in the examination of materials used in museums and collections as part of the Material-Checker project (MAT-CH), which aims to innovate the Oddy test and contribute to a higher standard of pollution control with improved precision, reproducibility and sustainability, by combining prototypes, performance and in-situ testing, as well as results from microscopy and instrumental analysis.

The second part of the book presents a number of different case studies. **Mastroiacovo et al.** present the results of the design and use of registration tables based on unified criteria for the preliminary stages of cultural heritage diagnosis, highlighting the advantages, limitations and areas for improvement identified after the first application experiences. The article by **Araújo et al.** presents the risk management strategies carried out at the Oswaldo Cruz Foundation in Rio de Janeiro (Brazil), a pilot experience of implementing risk management using the ABC method developed by the *International Centre for the Study of the Preservation and Restoration of Cultural Property* (ICCROM) and the *Canadian Conservation Institute* (CCI). **Aguiar et al.** present the collaboration between the *Portuguese Catholic University* (UCP) and the *Divisão Municipal de Museus* (DMM) in the city of Oporto to reorganise the municipal storage facilities and set up the process of transferring the collections to a permanent home in a requalified building. Interestingly, the paper presents the adjustments made due to the outbreak of the pandemic to adapt to such an unexpected context. **Gkinni** presents the environmental documentation initiated by the relocation of the National Library of Greece from Vallianeio (an imposing neoclassical building in the centre of Athens designed to passively control the indoor climate) to *the Stavros Niarchos Foundation Cultural Center* (SNFCC), the first major cultural project in Europe to achieve LEED Platinum certification, where active and passive technologies make it one of the most environmentally sustainable buildings in the world. **Cosaert et al.** present the *Resilient Storage* project, which aims to develop a protocol for Belgian museums to implement short-term energy saving strategies, reduce CO_2 emissions and improve conservation conditions. Finally, **Lorenzo-Cases et al.** discuss the challenges posed by the archaeo-palaeontological collection from the *Sierra de Atapuerca* sites (Burgos, Spain) at CENIEHR (National Research Centre on Human Evolution), as well as the risk management measures and preventive conservation strategies carried out to prevent the collection from dissociation and mechanical damage.

From this broad perspective, it is evident that this field of study is constantly evolving and that there is a growing interest in new developments that result in conservation solutions to be implemented by conservators-restorers, museum curators, facility managers, etc., in their routine conservation decisions. This is a clear indication of the growing dialogue between the different professional profiles involved in protecting cultural heritage but also of the need for specialised forums to share, discuss and move forward together.

Valencia, Spain   Ángel F. Perles-Ivars
Valencia, Spain   Laura Fuster-López
Eindhoven, The Netherlands   Emanuela Bosco

Contents

Environmental Monitoring

Developing Conservation-Focused Curriculum to Advance Analysis of Temperature and Relative Humidity Data

Vincent Laudato Beltran, Jeremy Linden, and Annelies Cosaert

Abstract The collection of temperature (T) and relative humidity (RH) data is a ubiquitous component of collection management, and plays a central role in understanding the preservation state of the collection, building envelope performance, climate control system operation, and loan policies. Such insights require that the data be analyzed and interpreted, with the results effectively communicated to stakeholders with varying levels of expertise. Analysis of T and RH data often relies on the native software from sensor/data logger producers. However, there exist several conservation-focused analysis tools that examine this data from complementary perspectives; their collective presentation as a 'toolkit' allows users to draw upon the individual strengths of each, but is hampered by limited awareness of the available tools and a lack of didactic support to foster their effective use. The Managing Collection Environments Initiative at the Getty Conservation Institute has developed a workshop curriculum that begins to address this gap by (a) establishing a baseline of knowledge on the fundamentals of data analysis and visualization, (b) presenting a survey of existing online T and RH analysis tools, and (c) demonstrating their application and interpretation for various case studies. To date, the authors have organized two virtual synchronous workshops, and plan to development a free asynchronous course.

V. L. Beltran (✉)
Getty Conservation Institute, 1200 Getty Center Drive, Suite 700, Los Angeles, CA 90049, USA
e-mail: vbeltran@getty.edu

J. Linden
Linden Preservation Services, 35 Meadowview Dr., Brockport, NY 14420, USA
e-mail: jeremy@lindenpreservation.com

A. Cosaert
Royal Institute for Cultural Heritage, Jubelpark 1, Parc du Cinquantenaire 1, 1000 Brussels, Belgium
e-mail: annelies.cosaert@kikirpa.be

© The Author(s) 2025   3
Á. F. Perles-Ivars et al. (eds.), *Collection Care*, Springer Proceedings in Archaeology and Heritage, https://doi.org/10.1007/978-3-031-85655-6_1

1   Introduction

The collection of temperature (T) and relative humidity (RH) data has become a ubiquitous component of collection management. This information plays a central role in understanding the preservation state of the collection, the performance of the building envelope in mitigating exterior conditions, the adjustment of set points for the mechanical climate control system, and the redefinition of loan policies. While other agents of deterioration may be more impactful, T and RH are important considerations when deciding upon management strategies that have an extended effect upon the collection environment and institutional resources. Such insights require that this RH data, which may be voluminous, be analyzed and interpreted, with the results effectively communicated to stakeholders with varying levels of expertise.

The analysis of T and RH data often relies on the native software from sensor/data logger producers. However, there exist several accessible conservation-focused analysis tools that examine this data from complementary perspectives, expanding the exploratory and narrative power of the dataset. The collective presentation of these tools as a 'toolkit' allows users to draw upon their individual strengths, but is hampered by limited awareness of the available tools and a lack of didactic support to foster their effective use.

The Managing Collection Environments (MCE) Initiative at the Getty Conservation Institute (GCI) has developed curriculum for a multi-session workshop that begins to address this gap. The first session establishes a baseline of knowledge on the fundamentals of data analysis and visualization, and proposes a template by which environmental analysis may be expanded. This is followed by a session on existing online T and RH analysis tools; described are their various numerical/visual outputs and means of assessing collection risk and facilitating stakeholder communication. The final session examines the application of these tools to a range of case studies and how data analysis can support decision-making on environmental strategies.

2   Session 1: Fundamentals of Analysis and Visualization

As the cultural heritage field increasingly collects data about the environment, there is a need to adequately analyze and interpret this wealth of information. While the immediate result is an improved understanding of object display and storage conditions, equally important is the subsequent communication of these findings to a diverse group of stakeholders responsible for environmental management, including directors, conservators, collection managers, registrars, curators, facilities and maintenance staff, conservation scientists, engineers, and architects. It is crucial to consider their varying levels of expertise with respect to data analysis to establish awareness of the environmental issues and support the decision-making process.

Since potential users of T and RH analysis tools may come from different academic backgrounds, access to training on the fundamentals of data analysis and visualization can establish a baseline of knowledge. The initial session of the workshop curriculum provides an overview of numeric analysis and visualization techniques; while these are by no means exhaustive, they offer a template for expanding environmental data analysis.

Note that the process of collecting environmental data is not discussed in depth, with the assumption that each participant is engaged in data collection. However, following collection of raw data, a data cleaning step is encouraged, as inconsistent data can result in erroneous conclusions. This involves the correction of inaccurate records—due to system maintenance (e.g., calibration), sensor malfunction, etc.—by replacement or deletion.

2.1 Numerical Analysis

The quantitative analysis of environmental data provides a framework by which it can be organized, analyzed, interpreted, and presented. The curriculum includes discussion of psychrometrics, numerical indices, and forecasting.

Psychrometrics refers to the physical and thermodynamic properties of gas–vapor mixtures, of which the most common is air and water vapor due to its relevance for air-conditioning or HVAC systems. Using concurrent T and RH data and a given elevation or barometric pressure, the user can further describe a parcel of air by calculating a suite of related psychrometric properties, including wet bulb temperature, enthalpy, specific volume, humidity ratio, and dew point temperature. The latter two properties describe the quantity of water in the air, and are particularly important for understanding the driving forces on RH, which is dependent on T and moisture content.

Various numerical indices can statistically characterize a dataset. The central region of a dataset is described by the average and median, while its dispersion can be depicted by the standard deviation and ranges (differences between key values such as maximum/minimum or 75th/25th percentiles). Transitory risks can be highlighted when applied to specific subsets, such as seasons, periods of changing HVAC operation, or notable occupancy or climatic events. Further, the analysis of discrete time windows (e.g., 24 h, 7 days) can smooth short-term variability (via a moving average, potentially relevant for massive objects that respond slowly to RH) or quantify short-term fluctuations (via a moving range).

When paired with models of object or building response, environmental data can be employed as a forecasting tool. Preservation metrics assess the biological, chemical, and mechanical risk to an object or collection, while building simulations predict energy use and the interior environment. Note that the calculations informing each projection may evolve with new research, and it is important that users understand their underlying rationale.

2.2 Visualization

The noted statistician, Edward Tufte, advocated a series of principles that support successful data visualizations, including showing data without distortion, presenting many numbers in a small space, encouraging comparison of different data, revealing data at multiple levels of detail, and serving a clear purpose (Tufte 2001). This workshop session embraced this philosophy as a means of better facilitating effective communication with stakeholders.

The collection of environmental data over time naturally lends itself to visualization as time series, which show time on the x-axis and variables of interest on the y-axis. The presentation of multiple variables or locations can be helpful in discerning trends in the dataset. For example, overlapping plots of T, RH, and dew point temperature at a single location can reveal temperature- or moisture-driven shifts in RH. Similarly, the combined visualization of different statistical treatments of a single variable (e.g., cleaned 15-min RH data, 7-day moving RH average, and 24-h moving RH range) and an environmental target zone on a time series highlights details that are masked when looking at the clean data.

Probability, which examines the likelihood of an event occurring, can be explored by cumulative relative frequency (CRF) or box plots. CRF plots display the variable of interest on the horizontal axis and the CRF (from 0 to 1) on the y-axis; if a variable value (e.g., T of 17 °C) has a CRF of 0.25 (equal to the 25th percentile), this indicates that 25% of the dataset is below that value. Assuming the dataset is relatively complete, CRF plots can visualize the percentage of time that data resides within a target zone.

Box plots provide a complementary view of probability by depicting the variable of interest on the x-axis and data groupings (e.g., locations, years) on the y-axis. The top and bottom of each box encompasses the interquartile range (defined by the 75th/25th percentiles) and whiskers show maximum/minimum values (or other percentile pairs such as 90th/10th) (Fig. 1). This graphical organization assists in the comparison of dispersion and skew in the data.

Visualization of T and RH data on a psychrometric chart illustrates its interdependence with moisture content. Commonly employed by HVAC engineers, the psychrometric chart shown in Fig. 2 is elevation-specific and depicts T on the x-axis, humidity ratio on the y-axis, and RH by isohume lines curving upward from left to right. (More complex versions show the full suite of psychrometric properties.) Multiple datasets and subsets can be plotted, with each point thermodynamically describing a parcel of air at a specific time. Further, the net direction on a psychrometric chart defines a psychrometric process: heating, cooling, humidification, or dehumidification. When compared to a target zone, the relative positioning of 'out-of-spec' data informs the psychrometric strategy needed to shift the air parcel closer to or within the target.

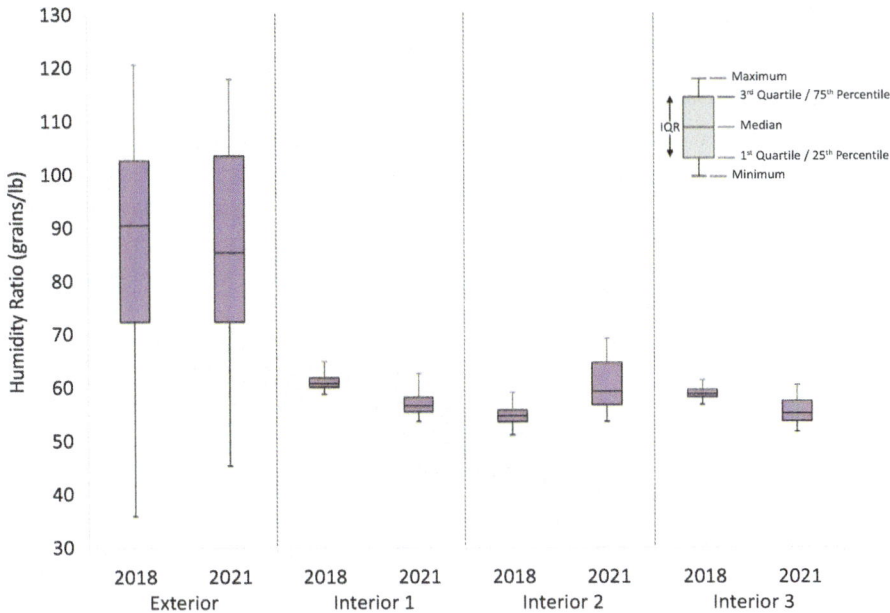

Fig. 1 Box plot of humidity ratio for the exterior and three interior locations during the summer of 2018 and 2021 (created with the GCI Excel Tools)

2.3 GCI Excel Tools

Developed for the GCI's MCE course 'Preserving Collections in the Age of Sustainability', the GCI Excel Tools makes accessible many of the aforementioned techniques. Due to its transparency and use of common software, this tool is presented as a didactic exercise at the close of the initial workshop session.

The GCI Excel Tools is composed of five modules:

- 'data_stats': Entry into the GCI Excel Tools, where the user inputs paired T and RH data and site elevation and various parameters and statistics are calculated. Subsequent modules are accessed in any order.
- 'time_series': Provides various templates of time series.
- 'crf' and 'boxplot': Provides templates for calculating and visualizing CRF and templates for creating box plots.
- 'psyc_chart': Uses T data and calculated humidity ratio values to overlay environmental data on an elevation-specific psychrometric chart.

The GCI Excel Tools has been largely limited thus far to participants in GCI courses and workshops, but are available upon request from the author and will soon be accessible online. Though developed as a teaching aid, this analysis and visualization tool has proven valuable in practice; users have reported increased

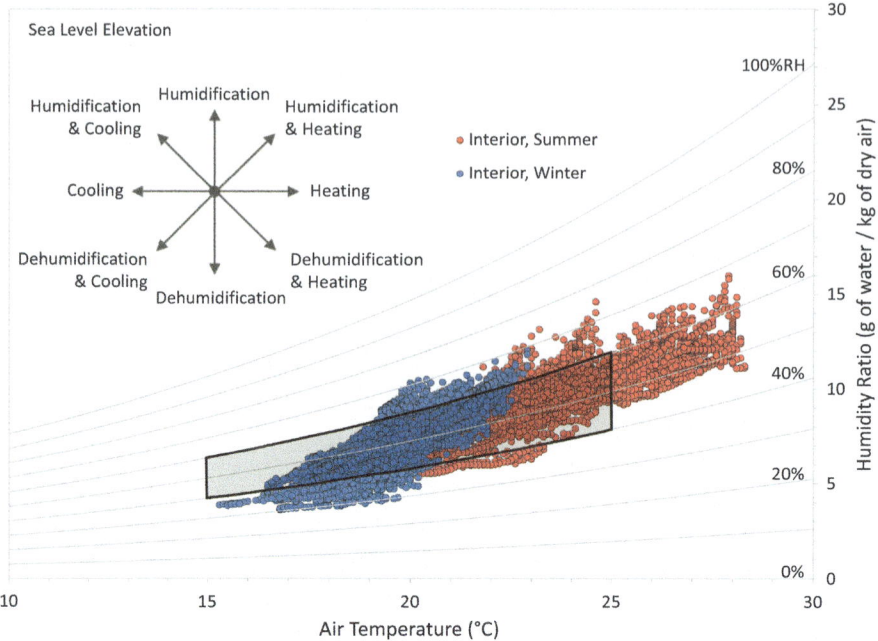

Fig. 2 Psychrometric chart showing seasonal interior data and a target zone (created with the GCI Excel Tools)

confidence in the data collection and analysis process, more nuanced communication with a range of stakeholders, and advanced access to the decision-making process.

3   Session 2: Online Tools Survey

T and RH analysis tools have the potential to support the environmental needs of collection management; a survey conducted by Cosaert and Beltran (2021) identified the following motivations:

- Redefinition of environmental set points to support more sustainable practice
- Comparison of environmental requirements based on collection composition and the existing interior climate
- Examination of potential consequences when moving objects between two different climates
- Analysis of display case performance.

However, research on preventive conservation tools by Katrakazis and Simon (2018) emphasized that learning a new tool is a strain for many collection care staff, particularly if supporting didactic material is not accessible.

Described in depth in a 2021 ICOM-CC paper by Cosaert and Beltran, the curriculum for the second workshop session focuses on a suite of T and RH analysis tools, including the GCI Excel Tools, eClimateNotebook (Image Permanence Institute, 2020), Physics of Monuments Online Application (Technical University Eindhoven; Smulders and Martens, 2014), HERIe (Jerzy Haber Institute; Działo et al., 2013) and the Calculator of Energy Use in Museums (Conservation Physics; Padfield, 2010). These specific tools were selected based on the following criteria:

- Carries out a clearly defined function
- Oriented towards conservation practice and implementation
- Assists in the decision-making process
- Free or relatively inexpensive (less than ~100 USD/year)
- Accessible online or via download.

The strengths of each T and RH analysis tool are highlighted by discussion of themes such as alignment with environmental standards, addressing risk, mechanical strategies and energy use, and communication. Also compared are the various functions of each tool, including the required data input and format, and the resulting data output and visualizations. The session closes with exercises applying these tools to given and user-provided datasets.

4   Session 3: Case Studies

The tools discussed thus far represent one step in the process of T and RH data analysis. Environmental data collection is now an accepted part of professional best-practice in cultural heritage. Due to the realities of limited budgets and staffing, decisions for preservation and operation must be data-driven and based on documentation for a specific institution. However, there often exists a gap between the gathering and eventual application of that information; for all that tools do, it can nonetheless be a struggle to understand what the data means in the context of the institutional setting.

Arriving at those decisions—the appropriate choices for a unique collection in a unique institutional setting—is a complex process that is the focus of the third workshop session. At the root of the issue is the deceptively simple question, 'What is the problem that needs to be solved?' Cultural heritage is moving away from prescriptive numbers-based standards and towards a range of acceptable options for institutions to consider. Preservation and operational choices must be based on institutional context: environmental conditions, collection needs, building envelope, climate management systems, and an understanding of how each relates to each other.

The analysis of T and RH data in the context of the decision-making process for preservation and sustainable building operation is, in part, an exercise in diagnostics—the process of identifying a problem by its nature and circumstances. The data provides an initial illustration of actual conditions versus potential goals;

using a series of algorithms and visualizations, the tools discussed in this workshop curriculum offer an initial analysis of the quantity and meaning of that difference.

However, data analysis—implying the application of science or mathematics on a dataset that provides a reproducible and quantitative result—alone does not provide full comprehension of a real-life system. The determination of what is 'best' with respect to collections and building preservation, sustainability (both in terms of energy use and organizational longevity), strategic planning, and daily operational activity necessitates effort beyond applying given T and RH set points.

Equally important to data analysis is the subsequent step of interpretation, incorporating factors of qualitative significance and of a potentially subjective nature. While use of similar T and RH data in the analysis tools will produce comparable mathematical results, the actual causes of, or impact resulting from, those environmental conditions and the appropriateness of that environment for a given situation will vary significantly depending on local circumstances.

The curriculum for this workshop session considers a range of factors for four case studies (Table 1). Though their relevance will vary between institutions, these factors generally align with the following categories:

- Geographic setting and local climate zone.
- Characteristics of the building or structure, including envelope design and material, floors and subgrade areas, and the location of collections within the space (e.g., interior or exterior walls).
- Collection type and significance: organic versus inorganic materials, varying object sensitivity to environmentally-driven risk, and the characterization of relative object significance to inform long-term retention.
- Capacity of mechanical environmental control systems, or the lack of a mechanical presence, and identification of zones—non-collection, mixed-use collections and people, unoccupied collections storage—served by the system.
- Preservation and sustainability goals, or other mandates or targets, whether internally or externally imposed (e.g., loan requirements, meeting defined preservation standards, energy consumption or carbon footprint objectives).

The collaborative interpretation step offers a holistic understanding of the institutional context to determine what factors may influence both the cause and meaning of T and RH conditions. Any perceived deficiencies in environmental performance may be solved or mitigated through intervention or correction of contextual factors—improved building envelope performance, improved collection housing, redefinition of appropriate environmental conditions—rather than solely focusing on adjustments to the mechanical climate control system.

Context is critical for environmental and preservation assessments, as is the recognition that preservation and sustainability goals are not simply quantitative targets but also encompass qualitative factors. The dual approach of data analysis and interpretation provides clarity of need and strategy for the modern management of preservation environments where there is no 'one size fits all' solution.

Table 1 Institutional context for the workshop case studies

Geography	Climate zone	Building	Collection	Climate control	Preservation goals	Sustainability goals
Southern coastal plain of large island, mountainous interior	Very Hot-Humid (1A)	Purpose-built storage vault, built 2010	Fine art, paper	T and RH control, shared zone with occupied spaces	Avoid mold growth, manage chemical decay, 70°F/50%RH	Reduce overall energy consumption through use of minimal outside air and reduced fan speeds
Desert alluvial plane surrounded by minor mountain ranges	Hot-Dry (2B)	Aircraft hangar built 1987	Mixed: aircraft, textiles, metals, plastics, rubber, paper	T control only	Control rates of corrosion and chemical decay	Manageable energy bills
Seaport city located on isthmus	Mixed-Marine (4C)	Storage warehouse renovated 2012 and 2019	Textiles	Purpose-designed T and RH control, dedicated zone	RH control, reduction of chemical decay rates, 60°F/45%RH with seasonal RH range	Minimal system operation using shutdowns, minimal fan speeds, demand-based operation
Small island separated from mainland	Cold-Humid (6A)	Renovated historic wooden structure	Fine art, paper	Purpose-designed T and RH control, dedicated zone	Avoid mold growth, seasonal T control	Variable seasonal operating parameters, reduce energy consumption (high energy cost on island)

5   Conclusion

The use of a 'toolkit' of T and RH analysis tools is complicated by the realities of limited experience and/or time by the collection care professional. However, data collection and analysis have become an important aspect of collection management and should be supported by accessible didactic material. The workshop curriculum detailed in this paper hopes to raise the floor of data analysis and interpretation, reduce the learning curve for individual tools, and promote dialogue among tool users.

In 2020, the GCI partnered with the American Institute for Conservation to conduct two virtual synchronous T and RH tools workshops during their Annual Meeting. The reliance on online tools was well-suited to a virtual platform and the extended workshop schedule of three two-hour sessions separated by one week provided ample opportunity for guided and self-exploration of the T and RH analysis tools and case studies. A voluntary discussion session was added to each workshop to address specific participant experiences. Both offerings were at full capacity and included 70 participants from eight countries.

While the authors remain open to additional synchronous workshops—a collaborative workshop with the Royal Institute for Cultural Heritage is planned for 2022— the didactic material is currently being adapted for an asynchronous course that would equitably expand educational outreach to the cultural heritage field and facilitate dissemination without the restrictions of time, place, and cost. An asynchronous framework will offer access to persons with varied experience levels and economic constraints, and from fields beyond conservation.

The tools presented in this curriculum will not remain static, and new tools are being developed to address specific user needs and take advantage of new technology. The potentially symbiotic relationship between tool developers and users was discussed by Cosaert and Beltran (2021), and lead to the GCI's organization of a 2019 meeting to discuss the collaborative development of preventive conservation tools. A GCI publication of this meeting report was produced in 2022, and highlighted the continued need for interdisciplinary dialogue that aligns ideas and strategies for data analysis with intuitive and easy-to-use tools.

Acknowledgements The authors are grateful for the support of the Getty Conservation Institute's Managing Collection Environments Initiative in advancing this work.

References

Cosaert, Annelies, and Vincent Laudato Beltran. 2021. Comparison of temperature and relative humidity analysis tools to address practitioner needs and improve decision making. In *Transcending boundaries: Integrated approaches to conservation. ICOM-CC 19th triennial conference preprints, Beijing, 17–21 May 2021*, edited by Janet Bridgland. Paris, International Council of Museums.

Cosaert, Annelies, Vincent Laudato Beltran, Geert Bauwens, Melissa King, Rebecca Napolitano, Bhavesh Shah, and Joelle Wickens. 2022. *Tools for the Analysis of Collection Environments: Lessons Learned and Future Development.* edited by Annelies Cosaert and Vincent Laudato Beltran. Los Angeles: Getty Conservation Institute.

Działo, A., M. Jędrychowski, Ł. Bratasz, L. Krzemień, A. Kupczak, R. Kozłowski, M. Łukomski, and Ł. Lasyk. 2013. HERIe: Quantitative assessment of risk of physical damage of cultural objects due to climate variations. Jerzy Haber Institute of Catalysis and Surface Chemistry Polish Academy of Sciences. https://herie.pl/. Accessed 2 Jan 2022.

Image Permanence Institute. 2020. eClimateNotebook. Rochester Institute of Technology (RIT). https://www.eclimatenotebook.com/. Accessed 2 Jan 2022.

Katrakazis, Theocharis, and Lambert Simon. 2018. *Survey of preventive conservation tool and resource users: Summary of findings.* Rome, Italy: Canadian Conservation Institute and the International Centre for the Study of the Preservation and Restoration of Cultural Property.

Padfield, Tim. 2010. Calculator for energy use in museums. https://www.conservationphysics.org/atmcalc/energyusecalc.html. Accessed 2 Jan 2022.

Smulders, H., and M. Martens. 2014. *Physics of monuments: Online applications.* Technical University of Eindhoven. http://www.monumenten.bwk.tue.nl/Algemeen/Applicaties.aspx. Accessed 2 Jan 2022.

Tufte, Edward R. 2001. *The visual display of quantitative information.* Cheshire, CT: Graphics Press.

The Human Sensor

Melissa King, M. Susan Barger, and Austin Senseman

Abstract There are thousands of mechanical sensors that can collect a great breadth of data in regards to collections care, however, there are still limitations on what data we can log from physical hardware. As humans, we are constantly observing the world around us, and if captured the right way, these observations have the potential to create even richer datasets. Conserv, a technology company focusing on preventive conservation tools, has incorporated an "observations" tool directly within our analytics software and mobile application. So far, our users have recorded observations having to do with the exterior weather, conservation wet cleaning treatments, building envelope issues, mechanical issues with their climate control system, and dew forming on surfaces among others. These observations can be visualized on an observation timeline alongside time series visualizations to provide more context when interpreting datasets. Users can also contribute observations with images directly from the mobile application, which is designed as a way to encourage more data entry from a larger team of people who are monitoring the collection, including gallery attendants and security staff. These observations allow users to better interpret their environmental data, but what if these observations were categorized and quantified in a way to more fully integrate into the analysis? We are building out the software to allow observations to be linked to specific objects and materials, provide users with the option of entering both quantitative and qualitative data, and assign categories for qualitative observations. The more data we can collect on the direct effect the environment has on our collections over time, the better we can supplement our understanding of environmental impact on materials through artificial aging techniques.

M. King (✉) · M. S. Barger · A. Senseman
Conserv Solutions, Inc., 5600 9th Ave, South Birmingham, AL 35212, USA
e-mail: melissa@conserv.io

© The Author(s) 2025
Á. F. Perles-Ivars et al. (eds.), *Collection Care*, Springer Proceedings in Archaeology and Heritage, https://doi.org/10.1007/978-3-031-85655-6_2

1 Introduction

In the heritage community, we have been using various instruments and sensors to monitor temperature, relative humidity, light, pollutants, and vibration for well over fifty years in our quest to better understand and control measurable factors affecting the longevity of our collections (Ntanos and Wei 2019). Our choice of sensors has progressed from cumbersome instruments like hygrothermographs and sling psychrometers to small, electronic data loggers that use sophisticated software and that can make multiple measurements at once. However, in spite of the advances in sensor and data logger technology, there are limitations in the range and subtlety of measurements that can be made. And, while instruments and sensors may record a large quantity of data, that data may not produce actionable information to aid in the care of collections, and we are missing key information about why the environment is doing what is represented in the data.

If an institution doesn't have a method to maintain and annotate data records, historical data may be lost along with any indication of what happened in the past related to temperature, relative humidity, light, and pests. Even fastidious record keepers may not keep an annotated log of their observations about their building, what is affecting their environment, or emergencies that are related to their sensors' records. For instance, they may not have an exact record of when a roof was damaged during a storm, or when there was a failure of a furnace or HVAC system, or when a special event occurred that had many people entering and exiting the building leaving the doors open. The memory of events may have only a hazy correspondence with the data recorded by their sensors.

With an easy method to tie events to the data from a logger or instrument, we can develop a more holistic and realistic understanding of what causes the changes in our environment. We would have a fuller dataset to apply to what we observe happening to collections. Standards of care for materials in museums have been developed based on the ten agents of deterioration defined by the Canadian Conservation Institute in the 1990s (Canadian Conservation Institute 2021; Waller 1994). We have worked to associate observed deterioration with scientific understanding to discern how cultural materials age and how they are specifically affected by various types of deterioration. Collecting institutions have objects that are undergoing natural aging all the time and we have not yet developed methodologies to easily connect the myriad of measurements we make in museum environments to the effects we observe in real time on objects. If we had an easy way to tie observations to our measurements and to analyze that combination of data, would it be possible to develop better and more sustainable methods for caring for collections?

2 History of Conserv

Conserv is a tech startup based out of Birmingham, Alabama, USA that set out to make data acquisition and analysis in cultural institutions easier and more comprehensible. Austin Senseman and Nathan McMinn, Conserv's founders, observed that the collection of data from data loggers was an onerous task that required going from logger to logger, downloading data and then making a spreadsheet to look at the data. This standard method generates lots of data, but it doesn't necessarily allow for real-time data analysis. It is cumbersome and time consuming—not something that leads to action, especially not immediate action. Using their backgrounds in data analysis (Senseman) and engineering (McMimm), the co-founders of Conserv looked for better answers to the environmental monitoring in cultural institutions.

To address the drawbacks from conventional data loggers, the Conserv team first developed freely available cloud-based software for better data analysis for all. Their software does not require people to learn to use Excel or other data management software, produces data that is easily interpreted, and generates reports that fit commonly needed formats, for instance a facility report needed to obtain loans. The free software has no limitations on the number of users within an account and how much data is imported from different dataloggers. The software includes analyses for temperature, relative humidity, dew point, mold susceptibility, light, all relative to each user's designated key performance indicators (KPIs). On the hardware side, the Conserv team also developed a sensor and gateway for environmental monitoring that uses wireless LoRaWAN technology, instead of Wi-Fi or Bluetooth transmission. Conserv sensors connected to the Conserv Cloud network can be read remotely via a mobile app.

3 Solutions

The observations feature within Conserv Cloud is accessible within the web and mobile application. When a user wants to record an observation, they are prompted to fill out a description of the observation, record the date and time it occurred or will occur, and they are asked to choose a scope (Location, Space, or Sensor) which then allows to specify which identifier within the scope. The observation feature also allows for users to upload images associated with an observation. There are two primary observation types: observations (general), and pest observations.

When choosing a time for an observation, users can record something in the past, present, or future. The observation can be an instantaneous moment or a span of time. Observations can also be made directly within the analytics view of the data as well by right clicking on a datapoint of the graph. Observations made about locations apply to all of the data within that location (i.e. a roof repair, HVAC problems, etc.). Observations about spaces apply to all of the data within that space, and then observations for sensors only apply to that specific sensor. Once observations are created, they appear in an "Observation Timeline" on the analytics page underneath the time-series graphs (Fig. 1).

Other groups have devised applications that enable members to record or report incidents and observations, but this capability has not been tied directly to monitoring environmental data. For instance, the LiveSafe® mobile application has been adapted by the Smithsonian Institution to allow museum workers to report and record events such as water leaks, housekeeping needs, mechanical and building issues, etc. to those who can address problems as they happen (Smithsonian Institution 2021). Conserv offers a similar capability, but it is not yet laid out with convenient categories or prompts encouraging users to record observations. However, it does allow users to directly annotate monitoring data with a timestamp and thus, observations/incidents are directly tied together in a way that assures that such connections will not be lost to the vagaries of memory. The ability to add observations with a mobile application with various users encourages on-the-go data entry from various stakeholders to increase overall data collection capacity.

4   How Have the Conserv Observations Been Used to Date?

Conserv software users have been able to record observations since the middle of 2019. As of December 2021, 189 users at 207 organizations have contributed 5,947 observations into Conserv Cloud. Of these observations, 84% of them are specifically related to integrated pest management. Since there was not a function to allow users to self-categorize within the software, the anonymized data had to be post-processed to aggregate within some general categories by searching for key terms associated with each category (for example, for weather, we used terms such as rain, sun, cloudy, etc.). It was difficult to categorize 46% of the observations, but of those that were categorized, it was a relatively close split between weather (16.5%), mechanical (17.9%), and building-related issues (14.2%).

Because Conserv Cloud allows infinite users within the account, we were interested in evaluating how organizations were benefiting from multiple users entering data in the form of observations, and how many observations organizations were making. So far, it seems that most organizations are adding 2–5 observations per month (Fig. 2a). Of the organizations that are adding observations, there are usually 1–3 users within the same account adding observations (Fig. 2b).

When observations are time stamped and linked to the rest of the data it allows users to better annotate their data and create a fuller understanding of their environment. In Conserv Cloud, users can indicate if an observation was in the past, the present, or in the future, and they also have the possibility of indicating that an observation spans over a period of time. From a data collection standpoint, it might be better if we can move closer to encouraging users to record observations "as they are happening." If users record data in real time it is more likely to be accurate than if it is a faded memory, and active data contribution will create a larger dataset. Of

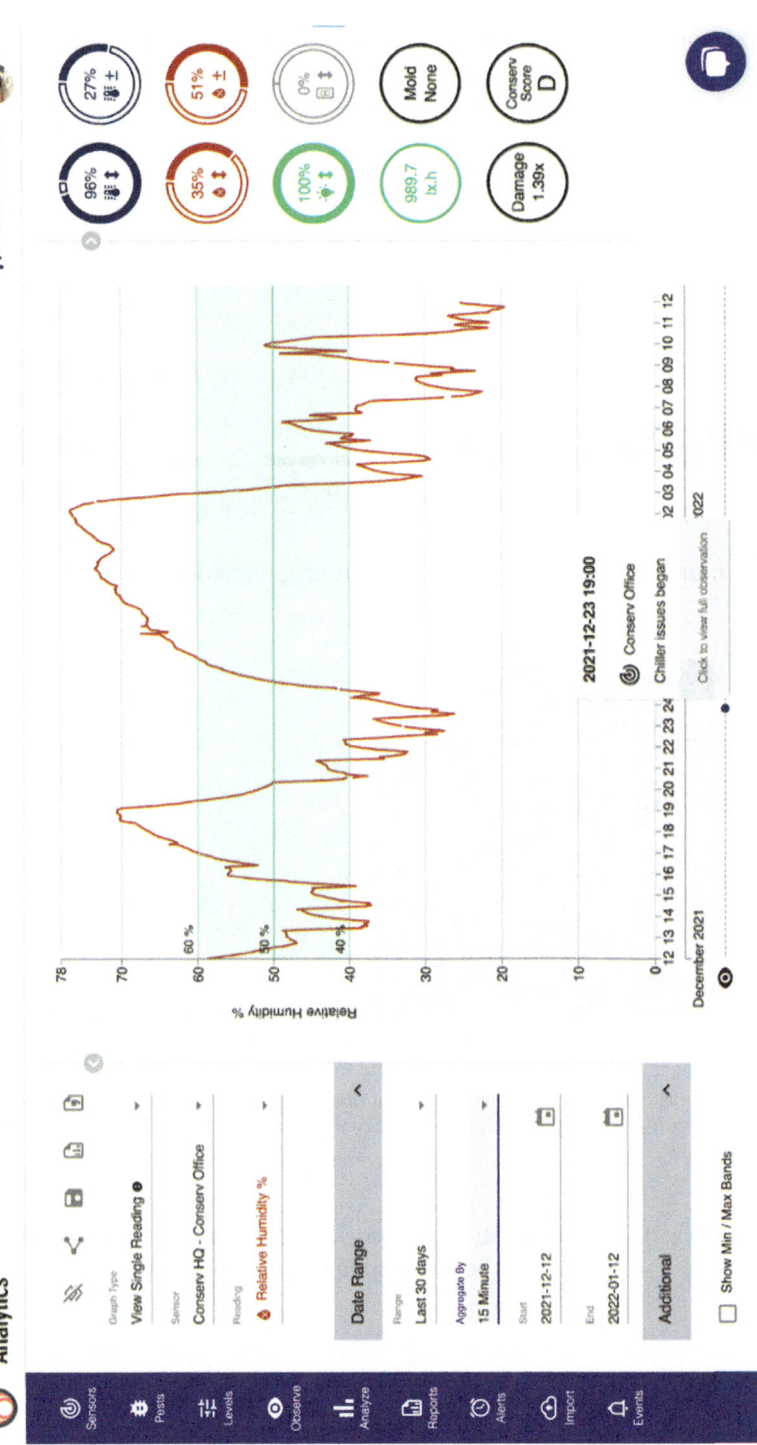

Fig. 1 A screenshot that shows the analytics feature in the software. Observations can be added directly to the graph by right-clicking on it. There is an "Observation Timeline" at the bottom underneath the X-axis that includes all of the observations made within the web and mobile versions of the software that show up as dots. When you hover over these with the mouse it tells you the observation that was made

a   How many observations are organizations making per month?

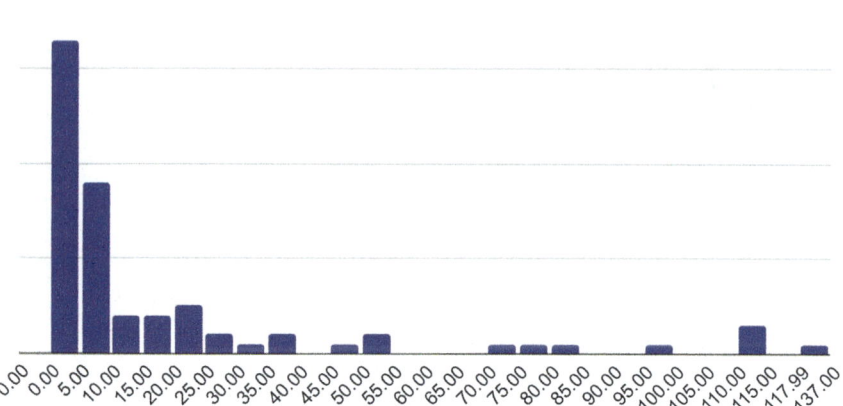

b   How many users are involved at each organization?

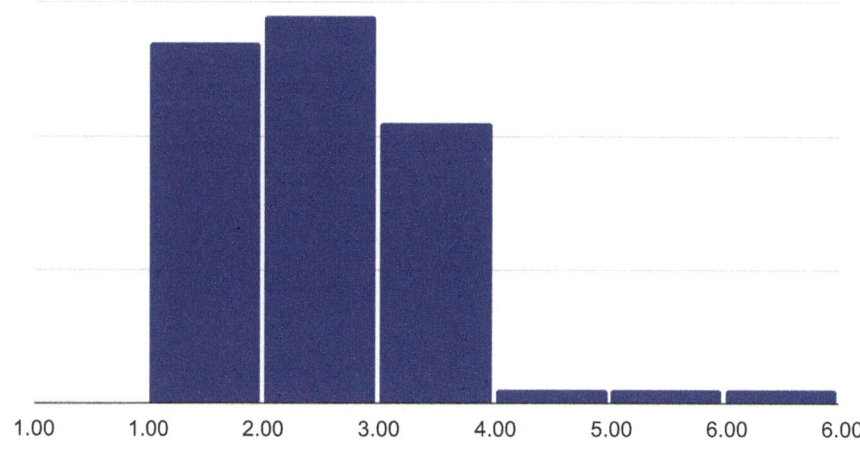

Fig. 2 Figure **a** is a histogram showing the number of observations individual organizations are making per month. Figure **b** is a graphical representation of the number of users adding observations within a single account

the user data analyzed, most observation categories are entered after the fact, except for "building issues" which seem to be recorded in real time (Fig. 3).

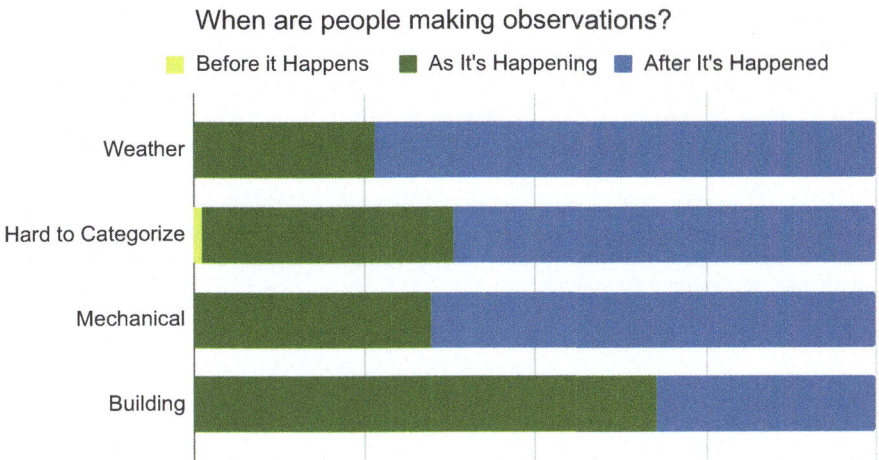

Fig. 3 This is a graphical representation when an observation is recorded as it relates to when it actually occurred. The data is broken down by observation category types

5   Conclusions and Future Work

At Conserv, it is our goal to make it as easy as possible for our users to record observations as they happen and to provide one place for such observations to be stored. We are working to develop ways to encourage more real time data entry by prompting users to assess events, observations, and data deviations. We have a multi-pronged approach of providing training and support to our users, by adding some pre-set prompt categories such as found in the LiveSafe® app used at the Smithsonian, and by continually improving our user interface in response to needs of our users. With defined categories for observations required upon data entry, there are more opportunities for future meta data analysis and improved analytical capabilities for users. We have already made it easy to add photographs of problems or pests and to annotate observations via the mobile app. Some early feedback about the observations has included proposed features that would allow users to filter observations by users and possibly including an approval process to address the accuracy of the data. It has been suggested that we incorporate an alert feature within observations if there are entries that require immediate attention such as a leak or environmental control issue.

Historically, the reason that cultural heritage institutions measure and keep track of their environmental conditions—temperature, relative humidity, light, pollution, pests and so on, is because such observations will lead to a richer and more complete understanding of how objects age and therefore, monitoring activities will lead to better care for collections. While collection professionals have been very good at collecting environmental data, this data has not always been used to improve object care. Conserv sees that the possibility of having observations directly tied to environmental data will provide an enhanced way to understand the problems of natural aging and will lead to better and more sustainable methods of collections care. Observations would lead to a fuller dataset to apply to what we observe happening to collections. What if we could also relate the ongoing effects of the outside weather on a building and its contents? Could we arrive at a deeper understanding of which and how quickly environmental changes cause damage to collections?

Acknowledgements The authors wish to thank their software users for contributing feedback about the observation tool and aiding in its development. They would also like to acknowledge the rest of the team at Conserv, particularly the product developers for interviewing users to design, produce, and update the tools within the software.

References

Canadian Conservation Institute. 2021. Agents of deterioration. https://www.canada.ca/en/conser vation-institute/services/agents-deterioration.html. Accessed December 2021

Ntanos, Konstantinos, and W. (Bill) Wei. 2019. Environmental monitoring. In *Preventive conservation: Collection storage*, eds. Isa lkin, Christopher A. Norris, 407–427. New York: Society for the Preservation of Natural History Collections.

Smithsonian Institution. 2021. Smithsonian institution climate change action plan. https://www.sus tainability.gov/pdfs/si-2021-cap.pdf. Accessed 1 December 2022

Waller, Robert. 1994. Conservation risk assessment: A strategy for managing resources for preventive conservation. https://www.museum-sos.org/docs/WallerOttawa1994.pdf. Accessed December 2021

Your Building Management Systems: A Treasure-Trove for Collection Care

Abed Haddad, Ana Martins, Richard Stomber, and Igor Rakush

Abstract The David Booth Conservation department and Building Operations department at the Museum of Modern Art (MoMA), in collaboration with the building automation firm TEC Systems, Inc., developed during the museum's last expansion a web based platform to monitor, evaluate, alert, and report on environmental conditions in gallery and non-gallery spaces, onsite storage, archives and libraries, conservation studios, workshops, and offices. The platform is connected to the Building Management System (BMS) and draws temperature and percent relative humidity (%RH) data from nearly 300 sensors across all art-containing spaces. TECConnector was developed as a web server on NodeJS and taps directly to the Enterprise Buildings Integrator (EBI) server to query data from historical trends and convert them into JSON formatted objects. This tool has considerably facilitated and enriched the dialogue between the conservation and operations teams, providing tools to better understand and act on the museum environment, while also bringing awareness to the needs of and requirements for collection care.

A. Haddad (✉) · A. Martins
The David Booth Conservation Department, The Museum of Modern Art, New York, NY, USA
e-mail: abed_haddad@moma.org

A. Martins
Van Gogh Museum, Amsterdam, The Netherlands

R. Stomber
Building Operations, The Museum of Modern Art, New York, NY, USA

I. Rakush
TEC Systems, Inc., Long Island City, NY, USA

© The Author(s) 2025   23
Á. F. Perles-Ivars et al. (eds.), *Collection Care*, Springer Proceedings in Archaeology and Heritage, https://doi.org/10.1007/978-3-031-85655-6_3

1   The Museum of Modern Art

1.1   Expansion History

From the outset, The Museum of Modern Art (MoMA) in New York City envisioned itself as a repository of the important developments in modern and contemporary art. Because of this ongoing commitment, the footprint of the museum has continually increased since its inception (Orlando 2020). MoMA was Initially hosted in rented quarters of a Manhattan office building on 57th Street in 1929, before settling into a five-story townhouse at the museum's current location three years later. Several expansions followed that dramatically increased the footprint of the museum. The first building, designed by Philip Goodwin and Edward Stone, which opened in 1939, tripled the exhibition space and added a film auditorium, a library, and two floors of office space. In 1951, a new wing by Philip Johnson was built along the museum's western edge, now the site of the residential Museum Tower; in 1964, another, larger Johnson addition appeared on its eastern flank. Further expansion designed by César Pelli, coinciding with the museum's 50th birthday added a new garden wing, a second film auditorium, multiple restaurants, and a bookstore. The museum expanded again for its 75th birthday. Yoshio Taniguchi's intervention included substantial renovations, a new atrium, and a new research complex housing a theater, classrooms, offices, along with the museum's Library and Archives. The most recent expansion, led by Diller Scofidio + Renfro in collaboration with Gensler, concluded in 2019 and increased gallery space by 30%.

1.2   HVAC at MoMA

Since MoMA found a permanent home on 53rd Street and for over 80 years, the standards for temperature and percent relative humidity (%RH) that ensure suitable exhibition or storage of objects have continuously evolved. Throughout, ducted central air conditioning has proved crucial for maintaining the desired setpoints; in fact, Garry Thomson in The Museum Environment goes as far as saying that "the complete answer to climate control is a central unit distributing fully conditioned air through ducts to all parts of the building (Thomson 1994)." In 1999, the addition of climate specifications for museums, galleries, archives and libraries in the Applications Handbook of the American Society of Heating, Refrigerating, and Air-Conditioning Engineers (ASHRAE) cemented the primary role air conditioning plays in environmental planning (ASHRAE 1999). That chapter of the Applications Handbook was last revised in 2019, which is the version currently consulted by MoMA (ASHRAE 2019).

The museum was at the forefront of this trend, with air conditioning first installed during the 1939 expansion in the Goodwin and Stone building. It is unclear whether the original heating, ventilation, and air conditioning (HVAC) system had stringent temperature and moisture controls. Later, in a press release from 1958, MoMA announced the arrival of a "23 ton Carrier Steam Absorption Refrigeration Machine [that] is about 7 feet wide [2.1 m] and 15 feet wide [4.5 m] to be installed at the museum (The Museum of Modern Art 1958)." This unit allowed the museum to maintain 24 hours of ideal temperature, relative humidity, and clean air throughout the entire building, and increased the footprint of air conditioning across its campus. This included: the frame shops, (un)packing spaces, on-site storage, restoration studio, and office spaces. Perceptibly, this unit was billed as revolutionary, and that "it will ensure proper atmospheric conditions for the Museum's Collection of paintings and for the hundreds of works loaned for temporary shows by other institutions and private collectors each year (The Museum of Modern Art 1958)." The release also mentions new ductwork, implying new fans; it omits piping, which may hint that the 1958 air systems were not multizone.

As the museum expanded, MoMA continued to incorporate steam absorption refrigeration through 2004, when it switched to electricity. The essential central components, chillers, cooling towers, and steam services were all at least 20 years old in 2004 and too small to support the expanded space. The cooling towers were located behind a masonry wall screen for aesthetics on the roof of the Goodwin and Stone building prior to 2004. They were subsequently relocated with the 2004 expansion: the cooling towers to the roof of the Taniguchi building and the chillers to the engine room.

Today, the HVAC system at MoMA encompasses a large and complex system of machinery and computer-controls. In its current iteration, it is made up of four electric centrifugal chillers totaling 3000 tons. The chillers are water cooled and the heat is rejected to the atmosphere using four rooftop cooling towers. The museum employs N + 1 redundancy on critical central HVAC equipment including chillers, cooling towers, and pumps. This redundancy ensures that equipment failure does not compromise the remaining assets' ability to maintain temperature and %RH setpoints (Nall 2015). MoMA uses "clean steam" supplied by Con Edison for heating and humidification. Conditioned air is supplied to galleries using approximately 100 air handling units ranging in size from 1,000 cfm to 25,000 cfm. Back of house spaces and offices that do not contain collections materials frequently do not employ humidity control.

2 Environmental Monitoring at MoMA

2.1 Institutional Guidelines

MoMA, like most medium to large institutions, relies on a Building Management System (BMS) to control and monitor environmental conditions across more than 69,000 m^2 (744,000 ft^2) of public and non-public spaces (Karolidis 2009). In its current state, the BMS consists of software and hardware; the software program runs on computer workstations and communicates with the hardware, which can be sensors or actuators needed for the control of building equipment for lighting, security, energy management, heating, ventilation and air conditioning, among other functions. At that time, the BMS at MoMA had been implemented by TEC Systems, Inc, which had managed the system since 2004 through the last expansion.

The ranges specified at MoMA have evolved alongside the ongoing discussion and scientific research on the subject, which is described at length elsewhere (AIC Wiki 2020). Research continues to inform decision making in relation to exhibition and storage practices, with a substantial body of research indicating that many materials are more resilient to environmental fluctuations than assumed previously (Nevin et al. 2018; Elkin and Norris 2019). Due to the large footprint and volume of visitors, MoMA arrived at moderately conservative and achievable facility ranges for exhibition and most storage, with relative humidity between 40 and 55 %RH with ± 5% maximum daily fluctuation and temperature between 19 and 25 °C (66–77 °F). Different ranges apply for library and archives as well as cool and cold storage for photographic materials. MoMA has also embraced the guidelines set forth by the International Group of Organizers of Large-Scale Exhibitions, known as the "Bizot Group," that stipulate Bizot guidelines for international loans of 40–60 %RH and 15–25 °C (59–77 °F) (Bizot Group 2009; Staniforth 2010). At the time of publication, the museum has begun work on upgrading to a new BMS and is looking to revise the setpoints for its galleries and art storage spaces.

2.2 Maintaining Setpoints

Environmental control falls primarily under the purview of the Building Operations department, where engineers and building managers actively monitor and act on the environment following guidelines established during building design and commissioning. These guidelines were informed by standards as those published by ASHRAE in 2019 (ASHRAE 2019), the conservation literature, and state code that governs ventilation. Historically, conservation and collection care staff have been involved in those initial planning stages, handing most monitoring duties to those in the Building Operations department. At a basic level, this can include monitoring of all temporary and permanent exhibition spaces, on- and off-site storage, offices, frame shops, imaging rooms etc.

Nowadays, continuous monitoring of environmental conditions warrants deeper involvement on the part of conservators and conservation scientists, with a focus on preventive conservation. For years, the conservation department at MoMA relied on sporadic and targeted temperature, relative humidity, and light monitoring for specific projects using stand-alone Onset HOBO U12 data loggers, and more recently, Onset HOBO MX1101 Bluetooth data loggers and Testo 160 THL Wi-Fi data loggers. PEM2 loggers were also used in storage spaces, both onsite and offsite, but have since been decommissioned in favor of data loggers from Conserv. All these loggers, however, can have logistical limitations. For example, Onset HOBO U12 loggers require downloading the data directly or using a HOBO U-Shuttle Data Transporter to upload it to proprietary software for interpretation. Granular data collection requires more energy and thus frequent change of batteries for all loggers. In other cases, the need for stable Wi-Fi or LoRaWAN signal can be an impediment. Deploying a dedicated network of environmental loggers throughout the museum, wireless or otherwise, is certainly feasible. Such a network was set up by the Conservation department to remotely monitor the museum's off-site storage facilities using Conserv data loggers and associated cloud-based monitoring platform. Setting up a similar network over the whole museum, however, would be very costly.

While it was possible to glean information from the BMS, requesting the data for periods of interest took a few days and subsequently needed to be processed into intelligible reports. Eventually, conservation gained read-only access to the BMS, but the platform is multi-pronged and not intuitive. Instead, the Conservation and Building Operations departments approached TEC Systems, Inc. about creating a customized interface capable of retrieving temperature and %RH data from the BMS sensors throughout the building, and that interface would also incorporate visualization and evaluation tools.

3 TEC Connector at MoMA

3.1 Development

Initial conversations between MoMA and TEC Systems, Inc. helped determine the kind of information needed to improve cooperation between conservation and building operations, namely granular temperature and %RH data from spaces exhibiting or housing collection objects. Discussions also touched on the environmental guidelines discussed above in addition to the ongoing methods employed for data acquisition and the current state of the environment at MoMA.

TECConnector platform was developed as an addition to the existing BMS, with a custom frontend developed and implemented by the graphics department of TEC Systems, Inc. (Fig. 1). The platform was developed based on ExpressJS, a free and open-source web application framework running on NodeJS JavaScript engine. The software is programmed in JavaScript, Python, and SQL programming languages. The platform culls from the same data BMS data collected using an Enterprise Buildings Integrator (EBI) server to investigate historic trends as well as real-time conditions, all of which are stored in a file-based database on the BMS server. Using an Open Database Connectivity (ODBC) protocol, the platform can connect to the BMS and run SQL queries to retrieve data. The system does not currently copy data into a separate database and relies entirely on the EBI server to directly query data for examination. TECConnector can then use this data for visualization, reporting, and out of range notifications.

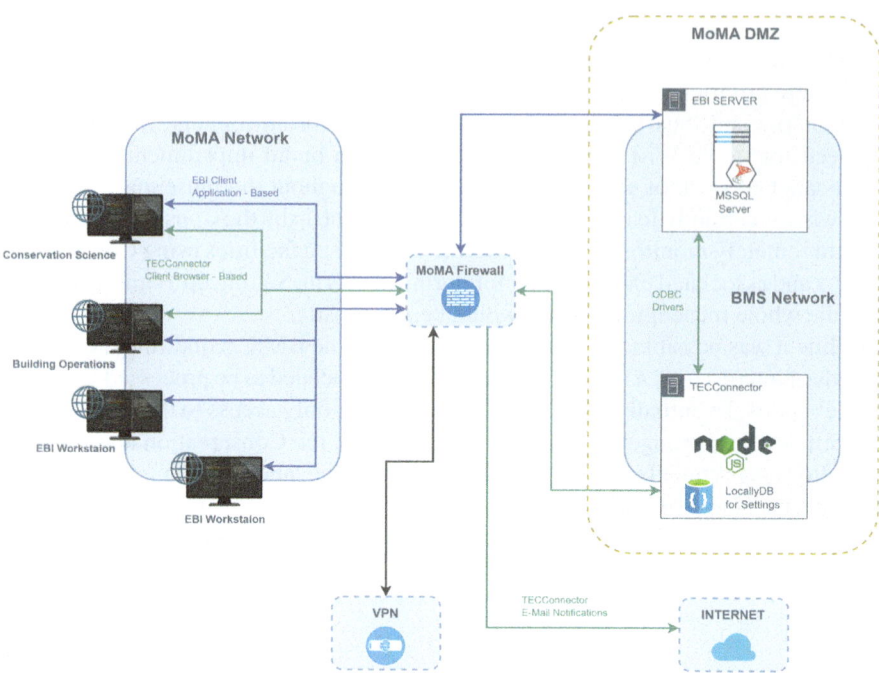

Fig. 1 A flowchart describing the backend communication between the BMS and MoMA networks

The system does not provide user login or specific authentication creden-tials since it was developed for internal use within the MoMA network. To access TECConnector, connection must be established to the museum's isolated local area network (LAN), which is managed and monitored by MoMA's Information Tech-nology department. MoMA's isolated BMS network strictly denies access without a deliberate connection on-site or remotely through a virtual private network (VPN). The platform relies on single traffic, read-only access of the EBI server, which is well protected within the demilitarized zone (DMZ) of MoMA's network. Therefore, the data called by TECConnector is safe within different layers of security, including read-only access and the DMZ envelope, which allows for visualization without user verification. However, the system is not without limitations; the loss of power hinders data collection by the sensors, and events that cause a system shutdown, such as fire alarms, can create unrecorded condition changes. Nevertheless, TECConnector had vastly improved day-to-day operations and tracking of environmental conditions at MoMA.

3.2 Implementation and Integration

During the 2004 Taniguichi expansion, most sensors were placed inside wall cavities as an aesthetic choice. But in an effort to better reflect gallery conditions, most sensors were refixed to permanent walls during the last expansion. By design, these sensors can be mobile but general preference is for permanent walls to ensure continuity of the data for year-on-year trends and historical patterns. After ensuring successful connection to the BMS, all the sensors were checked for accuracy by comparing readings from a handheld environmental monitor (ELSEC 765) at the launch of TECConnector.

The Conservation department tracks the temperature and relative humidity provided by nearly 300 sensors across all art spaces. These sensors include both humidistats and thermostats; some public locations like lobbies, corridors and stairwells are only monitored for temperature. The sensors chosen at the time of the 2019 expansion were Veris CWLSHTM sensors, which rely on a profiled thin-film capacitive humidistat for %RH and a 20,000 Ohm NTC resistive thermistor for temperature, and these sensors have and $\pm$ 1°F and $\pm$ 2%RH accuracy. Older spaces have ACI A/RH2-20 K-R2 sensors, which rely on a capacitive with a hydrophobic filter for %RH and a 20,000 Ohm resistive thermistor. Prior to 2004, most sensors were pneumatic. From the outset, all sensors were mapped out across galleries, lobbies and platforms, collection preparation and study rooms, on-site storage including cold and cool storage for photographs, frame shop, archives and library spaces. This network allows for both localized condition overviews (e.g. objects on loan in a specific gallery) and birds eye views across many sensors (e.g. by floor, exhibition, air handling unit, inter alia). At the time of this publication, work is ongoing to further upgrade these sensors across the museum.

3.3 Modules

There are several customizable modules (Fig. 2) that make up TECConnector: trends, reports, and key value control, i.e. out of range alarms. In the trends module, the data can be interactively plotted for any given period, going as far back as 2017. The values can be presented either as snapshot measurements at particular intervals, down to one minute, or as time-weighted averages (TWA) over a preset period: six minutes, one hour, eight hours, and 24 hours. The platform allows for overlaying multiple temperature and %RH trend lines, with the ability to toggle on and off any of the plotted sensors. Users can also create profiles that cull environmental data from preselected sensors. These profiles are highly customizable, and as a result, they allow for assessing the behavior of individual sensors, galleries, discrete HVAC units, entire floors, or even groups of objects across multiple locations belonging, for example, to a single lending institution. This also helps with documenting seasonal trends, identifying problematic galleries, sensors, or units.

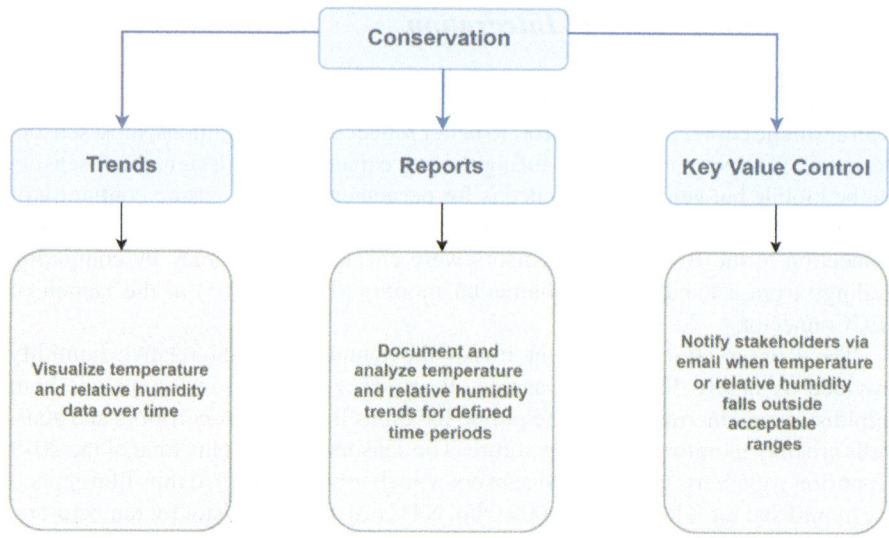

Fig. 2 The main modules that make up the TECConnector platform developed for MoMA

Beyond data visualization, the platform can also produce reports with performance analytics for a set profile of sensors as described above. The performance is evaluated per sensor by determining minima, maxima, means and standard deviations, in addition to the percentage of time spent outside specified temperature and %RH ranges, all over a chosen timespan. The units for temperature are easily converted between imperial (Fahrenheit) and metric (Celsius) in the reporting module for easier communication with lenders and partners outside the United States. The resulting reports are easy to understand by a wider audience that includes curators, registrars, art handlers, lenders, and others involved in the exhibition and movement of collection objects.

What is likely the most valuable feature of TECConnector is the email notification system, or key value control, which alerts pertinent stakeholders when temperature and %RH conditions are out of predetermined ranges over specified periods of time (Fig. 3). For example, while most notifications are only shared with the chief conservator, conservation scientists, and Building Operations, alerts to changing conditions in cool and cold photo storage are also sent to the photography conservator. These alarms are then addressed through investigation by Building Operations engineers with continuous updates on their findings, all collated in a chain under the original alarm email. These alerts also direct attention to malfunctioning sensors that require replacement if they are reporting faulty data or recording data at all. Sensors might also get relocated if inspection of the data points to a continuous bias in the reading, like being located next to a window or supply diffuser. The notifications can also be turned on and off, or have their frequency reduced, to prevent a massive influx of emails during scheduled maintenance events. Above all, the email alerts increased cooperation between Conservation and Operations departments and shortened response times to gallery condition issues or HVAC failures.

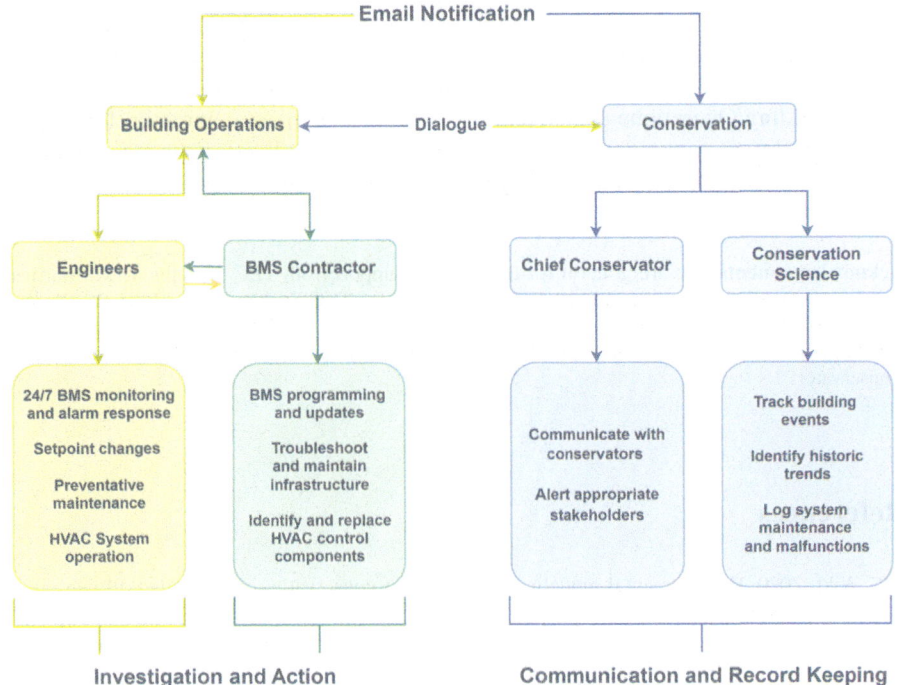

Fig. 3 The different types of responses by engineers, conservation staff, and BMS contractor to out of range email notifications at MoMA

Much like other modules in TECConnector, the alerts can also be customized. The set-up interface includes a free text field, which conservation scientists can use to keep track of building events or sensor issues. For example, some sensors are still entrenched within wall cavities near outside-facing windows and those often reflect closer to outside conditions over gallery ones; this is therefore noted as a reminder. Other information, such as fire department inspections, steam and HVAC shutdowns, maintenance, parts replacements, and extreme weather events are also logged by date.

4 Conclusion

The benefits gained from deploying TECConnector have exceeded simple monitoring, with significant cost and time savings through seizing existing live and past recorded data. Historical trends have painted a clearer picture of the performance of the mechanical systems and impact of seasonal changes. As the amount of data collected increases, wholesale assessments can be made and larger questions at the intersection sustainability and energy use can be investigated. While new avenues have been explored and implemented since TECConnetor, including infrastructure upgrades to the sensors described above, for several years the platform engendered fruitful collaboration and dialogue between the Conservation department, Building Operations teams, and the BMS management contractor. It brought together different viewpoints around the shared goal of collection management and laid the groundwork for continued engagement with external BMS contractors in the years that followed.

Acknowledgements We are grateful for the ongoing support from TEC Systems, LLC. Martins is indebted to Albert Hazan, head technician at MoMA for TEC Systems, Inc. (2004–2019). We are indebted to the work of the entire Building Operations team at MoMA. Special thanks to Jim Coddington, former Agnes Gund Chief Conservator, and Kate Lewis, current Agnes Gund Chief Conservator.

References

AIC Wiki. 2020. Environmental guidelines. https://www.conservation-wiki.com/wiki/Environmental_Guidelines. Accessed 21 Nov 2021.

American Society of Heating, Refrigerating, and Air-Conditioning Engineers (ASHRAE). 1999. *Museums, archives, and libraries*. Peachtree Corners, Georgia: ASHRAE.

American Society of Heating Refrigerating and Air-Conditioning Engineers (ASHRAE). 2019. *Museums, galleries, archives, and libraries*. Peachtree Corners, Georgia: ASHRAE.

Bizot Group. 2009. *NMDC guiding principles for reducing museums' carbon footprint*.

Elkin, Lisa, and C.A. Norris. 2019. *Preventive conservation: Collection storage*. Chicago, Illinois: Society for the Preservation of Natural History Collections.

Karolidis, Dimitrios. 2009. A report on building management systems in museums, with a reference to the archaeological museum of thessaloniki. In Paper presented at *BMS group meeting, Amsterdam*.

Nall, D.H. 2015. Rightsizing HVAC equipment. *ASHRAE Journal* 57: 48–52.

Nevin, Austin, J.H. Townsend, Matija Strlic, Nigel Blades, David Thickett, and J.K. Atkinson. 2018. *Preventive conservation: The state of the art: Contributions to the Turin congress, 10–14 September 2018*. London: The International Institute for Conservation (IIC).

Orlando, Jordan. 2020. The once and future MOMA. *The New Yorker*, January 26.

Staniforth, S. 2010. *Museums and environmental sustainability*. Prague: Bizot Group.

The Museum of Modern Art. 1958. *Museum to be closed for five days for installation of air conditioning unit. Press release*. New York, NY: The Museum of Modern Art.

Thomson, C. Garry. 1994. *The Museum environment. Butterworth-heinemann series in conservation and museology*. ed. John Ashurst, 308. London: Routledge.

Dynamic Indoor Climate Conditions for Sustainable Preventive Conservation

Rick Kramer, Marco Martens, and Edgar Neuhaus

Abstract Although standards and guidelines have evolved according to the paradigm shift 'from an ideal strict climate to an appropriate climate', e.g. by discriminating between short-term and long-term fluctuations and including seasonal changes, not much has changed in practice. Implementation of non-fixed setpoints for temperature and relative humidity is often a bridge too far or even simply impossible due to limited in-house knowledge at heritage institutions and technological limitations of building's climate control systems. In the 2010's, research at Eindhoven University of Technology has led to the concept of dynamic climate control in which controlled variations are used to realize an optimal course of temperature and relative humidity over the year respecting collection and comfort requirements. In 2020, the spin-off DYSECO was founded to further develop this concept to a fully automated self-adaptive form of dynamic climate control to enable heritage institutions to operate their air handling systems in a sustainable way for robust preventive conservation. All air handling units at the Hermitage Amsterdam museum employ the DYSECO system since mid-2020. This paper presents the energy savings obtained from dynamic climate control (data from 2020/2021) vs static climate control (data from 2014/2015). Results for the Winter, Spring and Summer seasons show substantial savings of 55%, 57% and 43% respectively.

Keywords Climate control · Preventive conservation · Energy efficiency · Sustainability

R. Kramer (✉) · M. Martens · E. Neuhaus
Department of the Built Environment, Eindhoven University of Technology, Eindhoven, The Netherlands
e-mail: r.p.kramer@tue.nl

DYSECO B.V, Rotterdam, The Netherlands

Á. F. Perles-Ivars et al. (eds.), *Collection Care*, Springer Proceedings in Archaeology and Heritage, https://doi.org/10.1007/978-3-031-85655-6_4

1 Introduction

In the nineteenth century heating systems were introduced in public buildings to (i) provide some level of thermal comfort in Winter (approximately 15 °C) and (ii) prevent too high humidity levels in Winter. Since this limited control of temperature only and not controlling RH, it became clear that the indoor climate of museums and other heritage institutes plays an important role in the preservation of the collection including, in the case of historic buildings, the interior and building itself. A wrong indoor climate might lead to biological, chemical and mechanical degradation. Biological degradation is mostly linked to moist conditions in which mould growth or wood rot can occur. This is mainly the case locally near uninsulated building components, e.g. single glazing in old window frames. Chemical degradation is the rate at which degradation takes place due to chemical processes within the materials. A higher temperature and/or a higher relative humidity will increase degradation and shorten life expectancy for objects. Finally, mechanical degradation is directly linked to fluctuations in temperature and relative humidity. Dimensional changes caused by fluctuations can lead to cracks.

As technology advanced in the twentieth century to full Heating Ventilation and Air-conditioning (HVAC), it became possible to control the indoor climate more strictly: temperature increased from 15 °C to 21 °C to provide better thermal comfort and relative humidity could be controlled more strictly around 50% RH aiming to mitigate climate induced degradation risks. Until 15 years ago the general notion was that the indoor climate could not be too strict. However, in the beginning of the twenty-first century, it became clear that maintaining strict indoor climate conditions all year round often leading to moisture damage to historic building structures, thermal discomfort of visitors (mostly in Summer), and excessive energy consumption. Hence, it is evident that strict indoor climate control hinders sustainability targets, but even more importantly, hinders a robust preservation practice. Interestingly, Martens (2012) demonstrated that the indoor climate in museums that employed strict indoor climate control was not far better in terms of preventive conservation quality compared to museums with less strict indoor climate control. In other words, current HVAC systems are not very effective in reaching a very strict indoor climate in both time and space.

1.1 *Towards an Appropriate Indoor Climate*

Mostly driven by sustainability awareness, the conservation community has been progressively reconsidering temperature and relative humidity requirements, changing from pursuing the ideal indoor climate (no fluctuations in T and RH) to pursuing an appropriate indoor climate (allowing some variations). Based on research over the last decades, e.g. (Luciani 2013; Martens 2012), and also following the commonly used indoor climate guidelines for museums, libraries and archives (e.g.

ASHRAE, ICOM, GCI, IPI) it is well known what the indoor climate should look like for appropriate preservation of the objects. ASHRAE presents indoor climate classes that provide ample flexibility for temperature and RH control such as seasonal adjustments and ranges for short term fluctuations (ASHRAE 2019). Ankersmit and Stappers (2016) have shown how to make sustainability part of the decision-making process for selecting an appropriate climate class. However, until the 2010s, the different pieces of the puzzle were not consolidated to make the term 'appropriate indoor climate' more concrete, for example: How to integrate collection and comfort requirements?; How to differentiate between short term and long term variations in a real-time control?; What is the energy efficiency associated with different climate classes?

Therefore, Kramer (2017) focused in his Ph.D. research on developing indoor climate control strategies for museums based on controlled variations. The underlying idea was to find the optimum course of temperature and relative humidity over time to achieve the lowest energy demand possible within boundaries defined by collection and thermal comfort requirements. Moreover, a vast energy saving potential was demonstrated for several climate classes, up to 63%, using both building simulations and measurements in real-life at the Hermitage Amsterdam museum (the Netherlands).

1.2 Technological Gap

However, while most museums are willing to employ dynamic setpoints and move along with the state-of-the-art, museum staff are simply not yet equipped with adequate tools. Figure 1 illustrates that the process of selecting a climate class, translating this class to setpoints for temperature and RH, evaluating indoor climate measurement data and accordingly adjusting the control parameters, largely remains a manual process. Engineers are capable of programming the BMS, but they do not know much about preservation of heritage objects and the associated climate requirements. Some museums' conservation specialists do experiment by changing the setpoints for temperature and relative humidity at regular intervals. Mostly this is limited to a few times per year manually lowering or raising the setpoints to match the seasons. A lot of knowledge is needed to do this and both the HVAC engineer and the conservation specialist need to team up. The 'appropriate indoor climate' is not easy to implement for many institutions and the manual process is very likely to result in suboptimal results.

1.3 DYSECO: Dynamic Setpoint Control

Technological developments are necessary to empower museum staff with more control over their indoor climate as current Building Management Systems lack

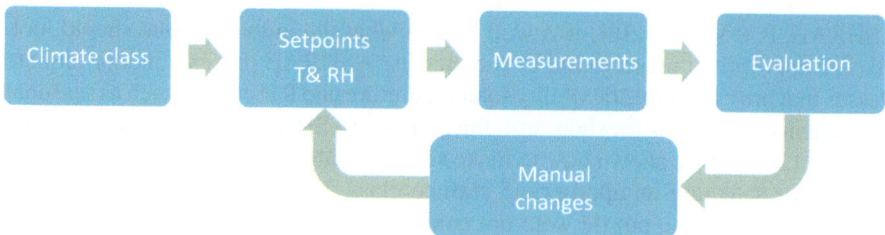

Fig. 1 Manual process of indoor climate control, monitoring and evaluation: Adjustments to setpoints are currently performed manually based on manual evaluation of climate monitoring data by museum staff

the flexibility of implementing temperature and humidity requirements that are specific for the heritage field, such as implementing dynamic indoor conditions (in which comfort and collection requirements are integrated). Hence, in 2019, the start-up DYSECO has been founded as a spin-off from Ph.D. research at TUE by Kramer (2017). Together with industry, the concept of dynamic indoor climate conditioning has been developed further into a control-module that can communicate with existing Building Management Systems. State-of-the-art knowledge on museum indoor climate control has been integrated into an algorithm that runs on the controller. For the first time, museums and archives can operate dynamic conditioning (e.g. conform ASHRAE class A1) in real-time: The algorithm on the controller calculates setpoints (upper limits and lower limits) for temperature and relative humidity on an hourly basis adhering to the boundary conditions set by the user, considering limits and permissible rates of change of temperature and relative humidity.

The controller is an add-on for any BMS system and overrides the fixed setpoints by the optimal dynamic setpoints. Figure 2 illustrates that only the selection of the climate class is a manual process and the rest of the process is automated. The controller receives indoor climate measurement data from the BMS, performs data analysis and calculates new setpoints and sends these new setpoints back to the BMS. These new setpoints are dynamic; every hour this procedure is repeated, to slowly and gradually adjust the setpoints over time ensuring an appropriate climate at the lowest energy demand possible.

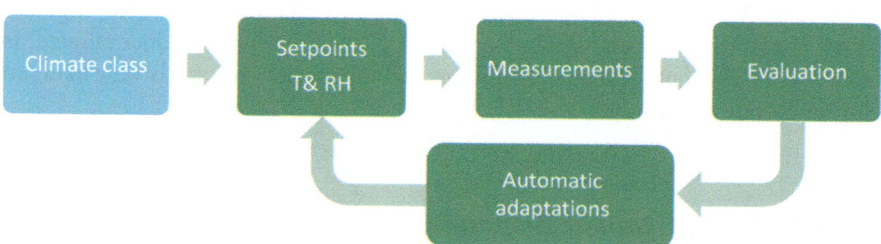

Fig. 2 Self adaptive dynamic climate control: The user selects the climate class manually and the rest of the process is automated

Since mid-2020, six DYSECO controllers were implemented at all six air handling units for the exhibition spaces of the Hermitage Amsterdam museum. Hence, since then, the museum has been running automatic dynamic climate control with small hourly adjustments and smooth controlled seasonal adjustments. The energy demand has been monitored since 2014. This paper compares the energy demand of 'strict control' and 'dynamic control' covering three seasons: Winter, Spring and Summer.

2 Methods

A comprehensive case study has been conducted at the Hermitage Amsterdam museum (Amsterdam, the Netherlands). The indoor climate and energy demand have been monitored since June 2014 until September 2021. For a comprehensive overview and explanation of the case study museum and the measurement campaign, we refer to a prior publication by Kramer et al. (2016). In summary, for the analysis in this paper, data of the following measurements at the air handling unit have been used (see Fig. 3): Electricity consumption of the steam humidifier and electricity consumption of the fan, water mass flow rate through the cooling coils and heating coils, water inlet temperature and water outlet temperature of cooling coils and heating coils. All measurements have been measured at a sampling rate of 30 s. From these measurements, the thermal power was calculated for heating, cooling, and dehumidification (deep-cooling).

The 'strict mode' data set, i.e. in which the indoor climate was maintained at 21°C and 50% RH, has been compiled from the following periods: December 2014–February 2015 (Winter), March–May 2015 (Spring), June–August 2015 (Summer). The 'dynamic mode' data set, i.e. in which the indoor climate was controlled via dynamic setpoints, has been compiled from the following periods: December 2020–February 2021 (Winter), March–May 2021 (Spring), June–August 2021 (Summer). In the dynamic mode, a custom climate class was employed which could be

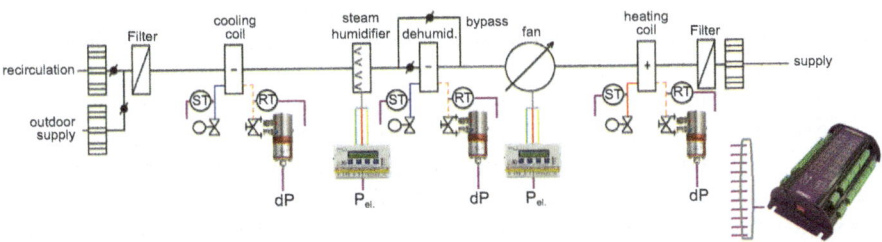

Fig. 3 Measurement setup of the main exhibition room's AHU. The data logger logged all signals (purple lines) at an interval of 30 s. Coil measurements included temperatures of the supply water (ST), return water (RT) and pressure drop of the water flow over the balancing valves (dP). Electric power consumption was measured of the steam humidifier and fan (Pel.). Figure adopted from (Kramer et al. 2016)

best described as ASHRAE class A1+: Absolute minimum RH = 40%, absolute maximum RH = 60%, permissible short-term RH-fluctuations = ± 5% (implemented as a maximum range between upper and lower limit), and the range was allowed to move between the absolute maximum and minimum RH at a rate of 5% per month. Moreover, absolute minimum T = 18 °C, absolute maximum T = 24 °C, permissible short-term T-fluctuations = ± 2 °C (implemented as a maximum range between upper and lower limit), and the range was allowed to move between the absolute maximum and minimum temperature at a rate of 2 °C per week.

The thermal energy demands as calculated from the measurements as depicted in Fig. 3, have been divided by generation and distribution process efficiencies to estimate the consumed electricity: the heat pump's COP of 4 for heating, EER of 3 for deep cooling, and an equivalent COP of 25 for high temperature cooling, directly from the Aquifer Thermal Energy Storage. Furthermore, the energy demand is presented as average electricity consumption per week in Winter, Spring and Summer.

3   Results and Discussion

Figure 4 shows the course of temperature (top) and relative air humidity (bottom) in February and March 2021 at the Hermitage Amsterdam museum in dynamic mode, i.e. controlling setpoints via DYSECO. As described in the Method section, the setpoints were automatically adjusted to meet ASHRAE A1+ requirements, i.e. ASHRAE class A1, but with stricter absolute limits. The ranges between the upper limit (green curve) and the lower limit (blue curve) were 4 °C and 10% RH. Within the range, the energetically most efficient setpoint was followed. In February, the outdoor temperature increased temporarily driving the indoor temperature range up, but a colder period followed driving the indoor temperature range down again reaching a minimum around 7 March. This shows the strength of adjusting the indoor climate using an algorithm instead of a predefined path of temperature and RH over the year: Depending on the case-specific internal and external loads, the indoor temperature and RH limits were smoothly adjusted to ensure compliance with collection requirements and minimize the energy demand. Small short-term fluctuations were allowed to prevent over-correcting the indoor climate and prevent constantly switching between heating and cooling. The actual indoor temperature (orange curve) fluctuated on a daily basis not more than 1.5 °C. The RH-range was not further adjusted in this period because the range was limited by its absolute minimum value of 40%. Interestingly, the RH was maintained in a very stable way with short-term fluctuations mostly below 1% RH. Only around 25 February, the RH was leaving the lower limit temporarily due to increased outdoor humidity levels, but short-term fluctuations safely remained between permissible limits.

Figure 5 shows the primary energy demand, in this case electricity, for air conditioning of the Hermitage Amsterdam's large exhibition space. The energy demand represents the average electricity consumption per week in Winter, Spring and Summer. The left-side bars represent the Reference situation, i.e. strict climate

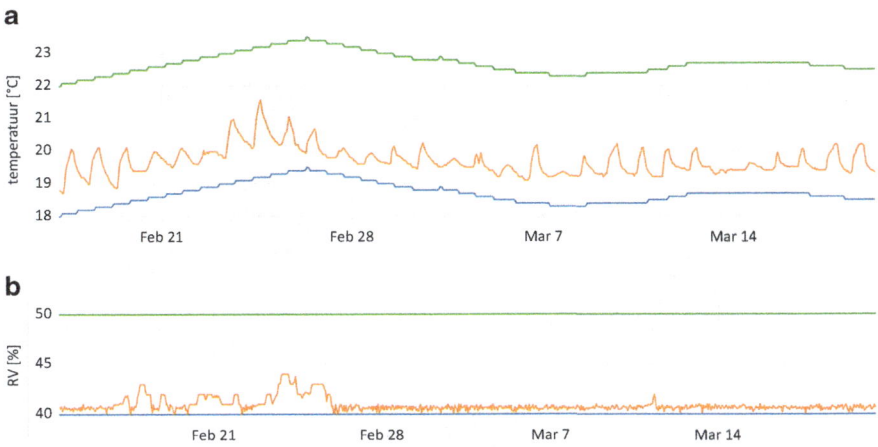

Fig. 4 Example of a dynamically controlled indoor climate in which the upper and lower limits followed collection needs according to ASHRAE class A1+ (i.e. minimum RH in winter of 40% and maximum RH in summer of 60%, with short-term fluctuations of ±5% RH) and 'standard level' comfort needs. Within the range, the energetically most efficient setpoint was followed. Top: Indoor temperature; Bottom: Relative air humidity

control at 21 °C and 50% RH, while the right-side bars show the energy demand under dynamic climate control.

In Winter, the energy usage was mainly determined by heating and humidification. The amount of electricity consumed by the heat pump for heating decreased by 75% in the dynamic mode. The amount of electricity consumed by the steam humidifier shows a 50% decrease in the dynamic mode. A lower temperature limit (absolute minimum of 18 °C instead of 21 °C) has led to significantly less heating and humidification, and the lower RH setpoint (40% instead of 50%) reduced the need for humidification even further.

In Spring, the dehumidification need is almost completely eliminated. Although dehumidification shouldn't be energy demanding in Spring, dehumidification actually formed a substantial share of the energy demand for air conditioning when fixed setpoints were employed. This was mainly due to frequent switching between humidification and dehumidification in Spring when the climate control aimed to maintain a very stable and strict indoor climate. Passive measures and passive buffering such as by hygroscopic building mass, have a negligible effect on the indoor climate when employing strict climate control, as the air handling unit control operates on a much smaller time scale. I.e., the air handling unit corrects the indoor climate actively, before the building materials are even able to desorb or absorb moisture from the air, which is generally a much slower process. Hence, dehumidification was often performed to remove moisture form the air that had been added by the steam humidifier just a few hours earlier. Also heating and cooling (dehumidification is realized by deep cooling and reheating) have been reduced dramatically. However, electricity consumed for steam humidification increased due to significantly dryer outdoor air

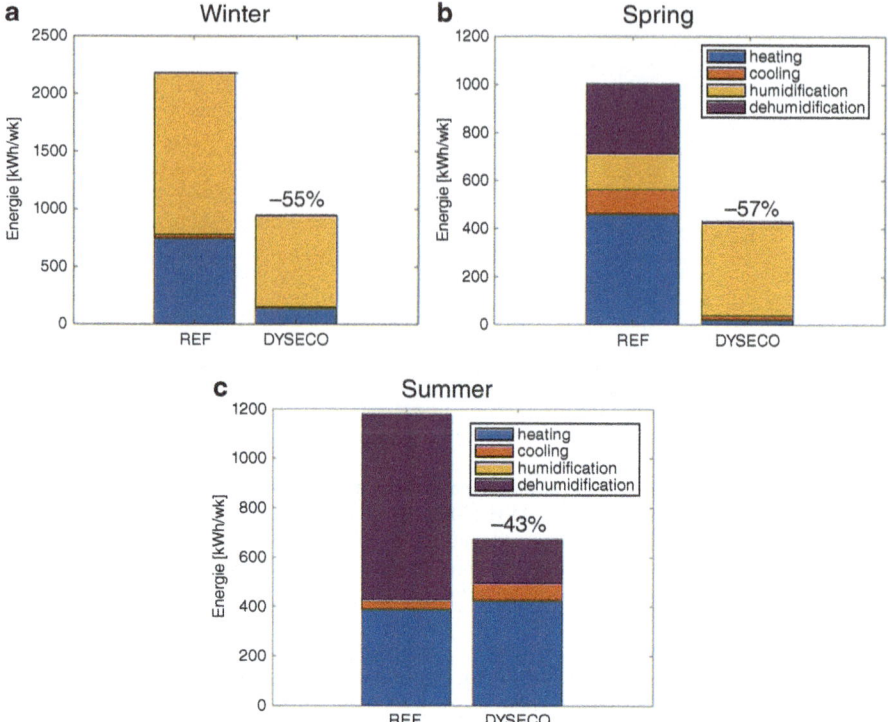

Fig. 5 Electricity per week used for heating, cooling and (de)humidification of the large exhibition space in Winter, Spring and Summer. The reference represents fixed setpoints of 21 °C and 50% RH, while the DYSECO scenario represents dynamic climate control complying to ASHRAE A1+

conditions in Spring 2021 compared to Spring 2014. This may be another reason for the reduced energy demand for dehumidification. In total, 57% of electricity has been saved in Spring.

In Summer, a significant reduction in electricity consumption by the reversible heat pump has been achieved due to a reduction in dehumidification demand.

4 Conclusions

Following the state-of-the-art in preventive conservation regarding optimized indoor climate conditions for cultural heritage requires a great deal of knowledge. Defining collection's and visitors' needs is a very important first step, but unfortunately an appropriate implementation is often a bridge too far or even simply impossible due to limitations of current climate controls.

From research conducted at Eindhoven University of Technology, DYSECO was founded as a spin-off company to further develop the concept of self-adaptive dynamic climate control and to enable museums and other heritage institutions to integrate preventive conservation into their building management systems' climate control.

From long-term in-situ measurements at the Hermitage Amsterdam museum, the Netherlands, substantial energy savings have been demonstrated in Winter, Spring and Summer, namely 55%, 57%, and 43% respectively. Undoubtedly, the demonstrated energy savings are prone to inaccuracies such as yearly differences in outdoor climate conditions, and specific savings will differ among museums. However, the results in this paper are promising and show that dynamic indoor climate control can significantly save energy while ensuring collection conservation and thermal comfort.

Acknowledgements The authors thank the Hermitage Amsterdam museum for their support over the years concerning data collection, climate interventions, and for their world-first implementation of automated dynamic climate control.

References

Ankersmit, Bart, and Marc HL. Stappers. 2016. *Managing indoor climate risks*. Cham: Springer.
ASHRAE. 2019. Museums, galleries, archives, and libraries. In *Heating, ventilating, and air-conditioning applications*, ed. by Mark S. Owen, 23.21–23.22. Atlanta: ASHRAE.
Kramer, Rick P. 2017. *Clever climate control for culture: Energy efficient indoor climate strategies for museums respecting collection preservation and thermal comfort of visitors*. Eindhoven: Eindhoven University of Technology.
Kramer, Rick P., H.L. Schellen, and Jos V. Schijndel. 2016. Impact of ASHRAE's museum climate classes on energy consumption and indoor climate fluctuations: Full-scale measurements in museum Hermitage Amsterdam. *Energy and Buildings* 130: 286–294.
Luciani, Andrea. 2013. *Historical perspectives on climate control strategies within museums and heritage buildings*. Italy: Politecnico di Milano
Martens, Marco HJ. 2012. *Climate risk assessment in museums: Degradation risks determined from temperature and relative humidity data*. Eindhoven: Eindhoven University of Technology.

Harnessing Data Science Technology for Environmental Monitoring at the V&A

Bhavesh Shah and Emily R Long

Abstract The Victoria and Albert Museum (V&A) has one of the world's largest museum environmental monitoring systems. Over the last 10 years, this system has collected data from over 500 sensors across several sites. This data has been processed, analysed and communicated using data science tools from R and RStudio. These tools have saved hours of data processing time, increased engagement with key stakeholders, and ensured the safety of the collections. This paper explores how these data science tools fit into the V&A's strategy for collections management. This paper will also introduce some methods of transforming, modelling, and translating environmental data into insights that can be communicated to museum stakeholders.

1 Introduction

The V&A has a collection of over 2.3 million art and design objects as a resource to inspire future artists and designers. The collection spans time, geographical regions, and materials, including furniture, textiles, photography, glass, paintings, and more. An object's chemical or physical properties determine how it interacts with the museum environment and whether the environment will result in deterioration. While individual object deterioration over time is too complicated to predict, monitoring the environment can prevent object damage. Therefore, collections care is increasingly dependent on the interpretation of environmental data.

Environmental data has been a critical part of recent research into monitoring heritage objects and sites. The environmental data can range from external climate data to interior gallery spaces, to microclimates like smart archive boxes with sensors (Gawade et al. 2021). Environmental data can be incorporated into the analysis of other datasets like visitor numbers, carbon dioxide levels in gallery spaces (Zhang

B. Shah (✉)
English Heritage, Ranger's House, Chesterfield Walk, London SE10 8QX, United Kingdom
e-mail: Bhavesh.Shah@english-heritage.org.uk

E. R. Long
Victoria and Albert Museum, Cromwell Road, London SW7 2RL, United Kingdom

© The Author(s) 2025   45
Á. F. Perles-Ivars et al. (eds.), *Collection Care*, Springer Proceedings in Archaeology and Heritage, https://doi.org/10.1007/978-3-031-85655-6_5

et al. 2021), or pollutants from archaeological woods (Rioual et al. 2021). Environmental monitoring data can also be combined with data on pests and mould growth (Kim et al. 2021). Statistical analysis of environmental data in combination with signals from the movement in cracked heritage buildings can be used to evaluate their structural health (Ceravolo et al. 2021). At the V&A, environmental data is used to evaluate present and future risks to objects and to define collections care strategies for future museum investments.

As technologies and tools for collecting data evolve, many museums are reflecting on their environmental monitoring systems. The V&A employs a system called OCEAN (Object Centred Environmental Analysis Network) to collect temperature, relative humidity (RH) and light data (Hancock 2004). The system was originally developed by Hanwell and the V&A, and it has been running for nearly 20 years. The OCEAN interactive software presents environmental data to museum staff and key stakeholders in the form of interactive maps, graphs, data summaries, reports, alarms and alerts. However, data science has developed newer and more powerful tools for processing environmental data. Thus, the coding language R and software RStudio have been incorporated into the V&A workflow to create reports with custom code (R Core Team 2021). While the V&A replaces the OCEAN hardware and software with a new environmental monitoring system (EMS), innovative research will continue to explore advanced methods of data science with R.

This paper will present a brief introduction to data science, and how that science needs to be communicated with key stakeholders. For example, two critical outputs of environmental monitoring are the monthly environmental report and the out-of-specification alarms which notify stakeholders when temperature and RH are outside an acceptable range. Then the paper will review some ways of transforming environmental data into risk metrics, and modelling the data through statistical and machine learning methods. After discussing data visualisation and communication through R-based web pages, this paper will conclude with a discussion of future projects that will utilise environmental data at the V&A.

2   Data Science

Data science is the application of statistics, coding, and expert knowledge to study and interpret data. The combination of these three components, as shown in Fig. 1, is what makes data science such an impactful tool for heritage and other fields. There is a wide range of data science tools and coding languages available to create bespoke tools for museum data. The V&A has used R and RStudio for the past 10 years to analyse environmental monitoring data. But why code in R instead of using more manual tools like Microsoft Excel?

Firstly, R can handle much larger datasets than Excel. For example, suppose 600 sensors across the V&A sites collect temperature and relative humidity data at 15-min intervals. This generates over 21 million rows of geospatial time-stamped data points per year or 1.75 million per month. In order to produce monthly summaries and

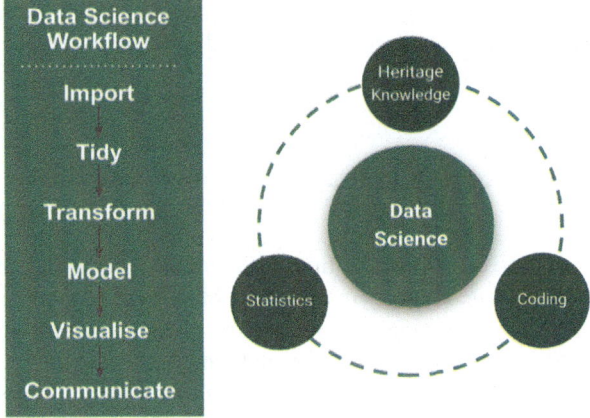

Fig. 1 A simplified data science workflow and a visualisation of the concept of data science as having three main components

environmental reports, a single data file would exceed Excel's limit of 1.048 million rows per sheet, but this amount is easily managed in R. Similarly, some environmental monitoring systems may output individual data files for each sensor. While small files are useful in the short term, R can be used to automatically combine these files for storage or more robust analysis. For example, in the process of archiving the data from the OCEAN system, R has processed over 225.5 million rows of temperature and relative humidity data from thousands of CSV files. Thus, a little code can handle a large amount of data.

Next, coding automates analysis so that it can be fast, repeatable, and reproducible (Wickham and Grolemund 2016). Over 14,000 packages are available in R to visualise and analyse data with simple functions, so code can be written quickly and executed efficiently (Wickham et al. 2019). By automating the creation of the V&A monthly environmental report, the report is created in only a few hours instead of a few days. Each month, the code can be run on a new dataset, and repeat the same analysis and report format. Finally, code can be shared amongst data scientists within an organisation, or more widely throughout the data science and heritage communities so that the analysis can be reproduced. Therefore, there are many benefits of coding for environmental data.

The data science process of analysing environmental data can be simplified into a workflow, as shown in Fig. 1. The first step involves importing the raw output data from the environmental monitoring system into R. Then, this data can be tidied into a standard format (Wickham 2014). Variables like sensor name, time, temperature, and relative humidity should be stored in separate columns. The time-stamped sensor readings should be stored in rows. In this format, the data is readable by both humans and machines. Then this data can be transformed to produce multiple outputs such as risk metric calculations and simple summary statistics. The data can then be modelled using statistical, machine learning, or psychrometric models. After visualising the outputs, the results can be communicated with the key stakeholders. Instead of focusing on the import and tidy stages of the workflow, this paper will discuss interesting areas of transformation and modelling for environmental data. However, the key to reaching the end of the data science workflow is to consider the interests of museum stakeholders.

3 Key Stakeholders

To effectively report the findings from environmental analysis, the data needs to be communicated clearly in an accessible format to a range of key stakeholders. At the V&A, the following stakeholders receive monthly environmental reports and alerts if the environment is out of specification.

First, the exhibitions team needs environmental reports to demonstrate that they meet the obligations of object lending agreements. If the exhibitions team receives an alert, they may need to notify lenders and collaborate with the estate team to bring the environment back into specification. The estate team manages areas that are conditioned by the Building Management System (BMS), which includes air conditioning, heating, cooling, humidification and dehumidification. In particular, the estate team needs alerts when environmental conditions may indicate a fault in the BMS. From the environmental summaries and out of specification information in the reports, the estate team will discuss any actions they have undertaken with the collections managers. Collections managers include curators, conservators and senior management who want to quickly understand the environment of objects in storage or on display in the museum. These stakeholders may choose to take preventive actions, such as removing waxes from certain galleries before the hot summer months. Finally, the monthly reports will include observations, investigations and actions that have been undertaken by the estate or collections teams to keep a record for auditing purposes. Therefore, there are many stakeholders who all need to understand the environment of the collections.

If the environmental reports can communicate effectively with stakeholders, they can have a large impact on decision making within the museum. Areas that are consistently out of specification may require future investment, such as fitting air conditioning systems or purchasing museum showcases. These projects require coordination with several teams, justifying and finding a budget, and project planning. Thus, the environmental data needs to be analysed to quantify and communicate the risk towards the collections.

4 Transformation: Risk Metrics

Once temperature and relative humidity data have been imported and tidied in RStudio, they should be used to calculate more meaningful values for stakeholders, like risk metrics. Risk metrics are useful to understand when and where risky events are occurring within the museum and if any management decisions are improving collections care. Risks can be broadly categorised into two types: one-time events that require immediate investigation and action, such as floods, or long-term risks that might require an interventive strategy.

Consider the long-term risk from a gallery that has intermittent yet consistent periods of high humidity. Many environmental monitoring systems can send alerts when the raw data conditions are outside the specified limits. However, with 96 data points per sensor being collected daily, these alerts can generate multiple alarms per day when conditions go in and out of specification. Staff can experience alarm fatigue and miss important alarms. Instead, calculating out-of-specification alarms as a percentage of the time the conditions are outside specification can help quantify long-term risks.

Next, consider short-term risks from extreme events. In these scenarios, damage models can be mapped against the stored and displayed collections to model the effects of the event (Strlič et al. 2013). For example, frost and dew point exceedances may occur when there are events with low temperature and high humidity. Mould calculations are useful when conditions are above 70% RH for several days. High mould scores require inspection of areas where this occurs or further interventive action. Lifetime or permanence calculations can be used to understand chemical degradation as a function of temperature and humidity (Schito et al. 2020). An example is the lifetime multiplier:

$$LM_i = \left(\frac{RH_{ref}}{RH_i}\right)^{1.3} . e^{\left[\frac{E_a}{R}\left(\frac{1}{T_i+273.15} - \frac{1}{T_{ref}+273.15}\right)\right]}$$

For time i, the lifetime multiplier (LM) compares each time-stamped temperature T_i and RH_i data point with the reference conditions T_{ref} (set to 20 °C) and RH_{ref} (set to 50% RH). The variable R is the ideal gas constant equal to 8.314 J/(mol K) and the activation energy E_a is 100 J/mol (Schito et al. 2020). Wood damage calculations are more involved and require specialist in-house developed software (Pretzel and Bridgland 2011) or uploading to the HERIe website (Kupczak et al. 2018). However, RStudio is a great tool for transforming environmental data by calculating these damage model metrics.

Change-point and anomaly detection methods are useful to show when conditions are behaving differently. For example, environmental data may be anomalous if the BMS has been re-programmed or an extreme climate event is occurring. The Seasonal Decomposition of Time Series by LOESS (STL) method (Bergmeir et al. 2016) can detect when air conditioning faults have occurred from environmental data. These metrics are useful for investigating whether damage has occurred to the collections, planning a course of action, and measuring the effectiveness of risk mitigation. Once these metrics have been assessed for effectiveness, they can be further tuned to reduce the number of alarms being sent by the system to highlight only critical events. The transformative capabilities of R also extend beyond risk metrics, damage models, and change-point methods to other kinds of modelling for collections care.

5   Modelling

The museum is developing methods of modelling how spaces behave to prepare for future scenarios. The future of the V&A includes the upcoming exhibitions programme, many gallery display projects, and new building projects, including two new V&A sites being built in East London. The planning needs to balance the requirements of collections care with other museum strategies, such as reducing energy usage for more sustainability, increasing visitor numbers, improving visitor comfort, and promoting staff engagement. Thus, the collections care team needs robust models for the effects of museum strategies on collections management.

The models can take a statistical or machine learning approach. Simple methods like linear regression can be accomplished by either approach (James et al. 2013). For example, linear regression can be used to model the temperature in internal building spaces using the external environmental conditions and time factors, such as season and month.

$$T_{internal_i} = \beta_0 + \beta_1 T_{external_i} + \beta_2 Season_i + \varepsilon_i$$

For a particular sensor at time i, this equation aims to explain the internal gallery temperature, $T_{internal_i}$, using the external temperature $T_{external_i}$ and the season $Season_i$, and accounting for a random error ε_i. The constant coefficients β_j of this model could be used to make seasonal adjustments to the BMS to reduce energy consumption. Linear regression models are relatively simple to model in R with the following code:

```
model<- lm(T_internal ~ T_external + season, data)
```

The 'lm' function stands for 'linear model' which is a part of the R stats package (James et al. 2013). There are many more machine learning models in R packages that can be easily used on tidied data (Kuhn and Wickham 2020).

There is an abundance of statistical methods that can be used within the R ecosystem. Even simple statistics, such as means and percentages, are useful to understand how the environment at different spaces and times compare. A fundamental statistical concept is linear correlation, which aims to quantify the strength of the relationship between two variables (Ceravolo et al. 2021). For example, if the temperature readings of sensors within the same gallery spaces are highly correlated, then a smaller set of sensor readings could be used to model the environment. Clustering algorithms are a group of more advanced statistical techniques that can be used to find subgroups within a data set (James et al. 2013). For example, Fig. 2 shows clusters of galleries from V&A South Kensington as leaves on a tree. If gallery spaces are on the same smaller branch, then they have similar environments, whereas galleries on different branches have different environments. Clustering and correlations provide a useful way of exploring the relationship between the environment in gallery spaces.

Different models should be used according to the amount and type of data, complexity and processing power required. Adding complexity to the models often

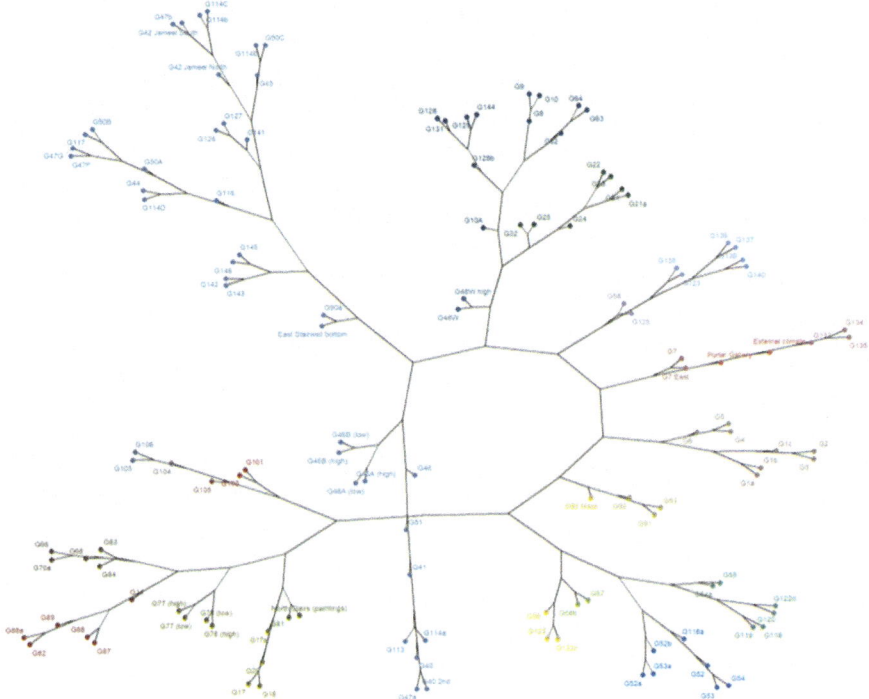

Fig. 2 Clustering analysis of the galleries at V&A South Kensington. The leaves of the branches represent the galleries

makes it harder to interpret and explain the outputs to key stakeholders. However, carefully selecting algorithms can provide targeted insights and balance the museum strategies. Although very useful, machine learning and statistical models generally require large volumes of data and computing power to fit the models. However, where there is plenty of environmental data available, these methods are useful to highlight critical and slowly developing issues.

Work is being conducted in collaboration with the estate team to see if modelling can provide improvements to reduce the energy demands of the museum in future climate scenarios. This combines the Met Office's future climate projection data from the HadCM3 climate model with the V&A's environmental data (Pope et al. 2000). Figure 3 shows the results of a psychrometric engineering algorithm that calculates the heating–cooling or humidification-dehumidification required to bring environmental conditions back into specification (Oughton and Hodkinson 2008). Additionally, a decision system is being developed with the estate team to avoid actions that are contradictory to the museum strategy. For example, cooling in the summer creates a more pleasant climate for visitors and staff. However, excessive cooling can be damaging to objects due to increased relative humidity. Modelling

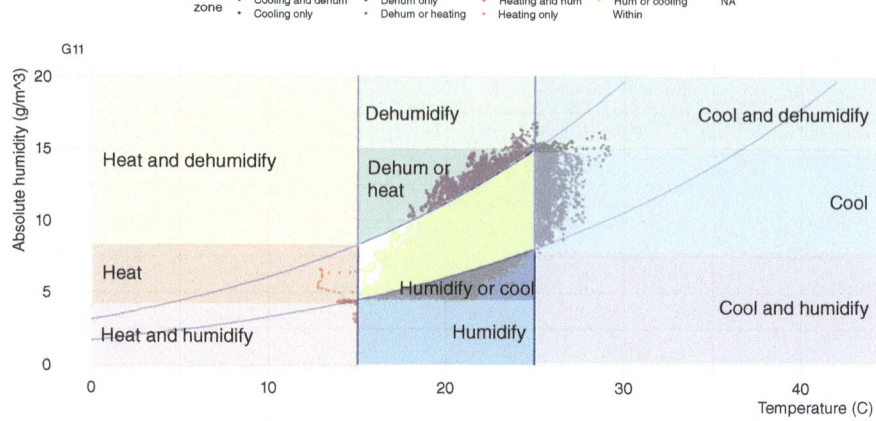

Fig. 3 A plot of the psychrometric algorithm results with V&A environmental data for South Kensington, gallery 11. Annotations indicate the actions that should be taken to return the conditions back to specification

adjustments to the museum environment by humidification-dehumidification and heating–cooling allows us to estimate which investments are required for the future.

6   Communication

The communication of large volumes of data creates the issue of how to filter out the noise and focus on the critical issues or results. Data visualisations like graphs and maps can make data more relatable for stakeholders. For example, the map in Fig. 4 visualises the variation in temperature across a simple map of the V&A South Kensington site. Visualisations should also be formatted so that the data tells a story (Grant 2019). Using the Shiny package in R (Chang et al. 2021), risk metrics and models can be presented in an interactive website with easy-to-use dashboards and downloadable reports. These web pages are accessible and engaging for key stakeholders. The R package leaflet (Cheng et al. 2021) can help transform Fig. 4 into an interactive map where stakeholders could see the historic gallery environments and the predictions for future climate scenarios. The presentation and communication of data to key stakeholders are being carefully considered as the V&A upgrades to the new EMS.

Temperatures in V&A South Kensington Galleries

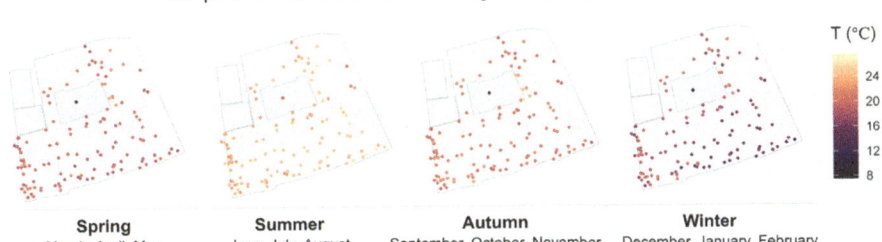

Fig. 4 V&A South Kensington average gallery temperatures mapped onto an interactive map using the R package leaflet (Cheng et al. 2021)

7 Future Development

The V&A is transitioning to a new environmental monitoring system. Since environmental data is being used in critical monthly and yearly reports, it is important that data from OCEAN is properly archived and accessible in a SQL database. The key to joining data from the two systems is metadata like the geospatial location of sensors, installation and calibration dates, and serial and identification numbers. The EMS team is also collaborating with the estates and collections teams to ensure that any metadata for space names and case numbers is being included in the EMS database. This metadata work will allow future environmental data projects to better connect to the BMS, Collections Management Systems (CMS), emergency response databases, museum maps, external weather data, future climate projections, Integrated Pest Management (IPM), and pollutant data.

When this metadata is combined with the OCEAN and EMS data streams, they form a wider data architecture for environmental monitoring at the V&A. With the foundation of a future-proofed data architecture, data engineering work will create a central database of the connected data streams. Then data science tools like Microsoft Azure and Power BI can automate the process of applying statistical and machine learning models without the need to manually download data. These platforms also allow R code to be combined for the development of systems that can analyse data in real-time. After some more development, the R functions coded at the V&A over the last few years could be formed into an R package and shared with the wider conservation community.

Training the preventive conservation community in using and developing data science tools is going to be increasingly important for the future. Courses like University College London's Data Science for Cultural Heritage MSc are addressing this issue by creating the first generation of data scientists specifically trained for the heritage community. Online courses in coding, statistics, and data science are also available in either paid or free formats (Cosaert and Beltran 2021). Many coders using data science tools rely on the support of a community. A group called ConCode (Conservators Coding) was established as a space for the conservation community to discuss and learn data science techniques in a friendly and supportive environment. As coding can seem intimidating, the space is designed to make the process as open and inclusive as possible (ConCode 2021).

8 Conclusion

As the V&A upgrades from OCEAN to EMS, it is important to evaluate the data science tools that have enabled environmental monitoring for collections care. This paper briefly introduced the data science workflow with an emphasis on transforming the data into risk metrics and modelling with statistical, machine learning, and psychrometric approaches. A crucial part of data science is to communicate the results with the key stakeholders within the museum to aid their decision making. The packages available in RStudio have reduced the effort required to analyse data, provided a clearer picture of risky areas in the museum, and enabled the beginning of research into the impact of climate change on the V&A. Together, heritage experts and data scientists can make data science accessible to everyone who has a computer, an environmental dataset, a bit of time, and a problem to solve.

References

Bergmeir, Christoph, Rob J. Hyndman, and José M. Benítez. 2016. Bagging exponential smoothing methods using STL decomposition and Box-Cox transformation. *International Journal of Forecasting* 32: 303–312.

Ceravolo, Rosario, G. Coletta, G. Miraglia, and F. Palma. 2021. Statistical correlation between environmental time series and data from long-term monitoring of buildings. *Mechanical Systems and Signal Processing* 152: 107460.

Chang, Winston, Joe Cheng, J. J. Allaire, Carson Sievert, Barret Schloerke, Yihui Xie, Jeff Allen, Jonathan McPherson, Alan Dipert, Barbara Borges, Mark Otto, Jacob Thornton, Prem N. Khan, Victor Tsaran, Dennis Lembree, Srinivasu Chakravarthula, Cathy O'Connor, Stefan Petre, Andrew Rowls, Brian Reavis, Salmen Bejaoui, Denis Ineshin, Sami Samhuri, John Fraser, John Gruber, Ivan Sagalaev, and R Core Team. 2021. Shiny: Web application framework for R. https://CRAN.R-project.org/package=shiny. Accessed 9 Jan 2022.

Cheng, Joe, Barret Schloerke, Bhaskar Karambelkar, Yihui Xie, Hadley Wickham, Kenton Russell, Kent Johnson, Vladimir Agafonkin, Brandon Copeland, Joerg Dietrich, Benjamin Becquet, A. S. Norkart, L. Voogdt, Daniel Montague, A. B. Kartena, Robert Kajic, and Michael Bostock. 2021. Leaflet: Create interactive web maps with the JavaScript 'leaflet' library. https://CRAN.R-project.org/package=leaflet. Accessed 9 Jan 2022.

ConCode. 2021. ConCode: A collaborative network: Coding for cultural heritage. https://www.concode.info/. Accessed 9 Jan 2022.

Cosaert, Annelies, and Vincent Beltran. 2021. Comparison of temperature and relative humidity analysis tools to address practitioner needs and improve decision-making. In *ICOM-CC 19th triennial conference preprints*, 1–11. Beijing: ICOM-CC.

Gawade, Dinesh R., Steffen Ziemann, Sanjeev Kumar, Daniela Iacopino, Marco Belcastro, Davide Alfieri, Katharina Schuhmann, Manfred Anders, Melusine Pigeon, John Barton, Brendan

O'Flynn, and John L. Buckley. 2021. A smart archive box for museum artifact monitoring using battery-less temperature and humidity sensing. *Sensors* 21: 4903.

Grant, Robert. 2019. Interactivity. In *Data visualization: Charts, maps, and interactive graphics*, ed. Grant Robert, 181–187. Florida, FL: CRC Press.

Hancock, Martin. 2004. The OCEAN project at the V&A. *V&a Conservation Journal* 46: 20–21.

James, Gareth, Daniela Witten, Trevor Hastie, and Robert Tibshirani. 2013. Linear regression. In *An introduction to statistical learning: With applications in R*, ed. James Gareth, Witten Daniela, Hastie Trevor, and Tibshirani Robert, 59–126. New York, NY: Springer.

Kim, Si H., Hyun J. Lee, Seon H. Jeong, and Yong J. Chung. 2021. Biological distribution and environmental monitoring for the conservation of Janggyeong panjeon depositories and daejanggyeongpan (Printing woodblocks of the Tripitaka Koreana) of Haeinsa temple in Korea. *International Biodeterioration & Biodegradation* 156: 105131.

Kuhn, Max, and Hadley Wickham. 2020. Tidymodels: A collection of packages for modeling and machine learning using tidyverse principles. https://www.tidymodels.org. Accessed 7 Jan 2022.

Kupczak, Arkadiusz, Mariusz Jędrychowski, Marcin Strojecki, Leszek Krzemień, Łukasz Bratasz, Michał Łukomski, and Roman Kozłowski. 2018. HERIe: A web-based decision-supporting tool for assessing risk of physical damage using various failure criteria. *Studies in Conservation* 63: 151–155.

Oughton, Dough, and Steve Hodkinson. 2008. *Faber Kell's heating air-conditioning of buildings*. Jordan Hill: CRC Press.

Pope, V.D., M.L. Gallani, P.R. Rowntree, and R.A. Stratton. 2000. The impact of new physical parametrizations in the Hadley Centre climate model: HadAM3. *Climate Dynamics* 16: 123–146.

Pretzel, Boris, and J. Bridgland. 2011. Predicting risks to artefacts from indoor climates. In *Proceedings of the ICOM-CC 16th triennial conference Lisbon preprints*, 19–23. Lisbon: Critério Artes Gráficas, Lda.; ICOM Committee for Conservation.

R Core Team. 2021. R: A language and environment for statistical computing. https://www.R-project.org/. Accessed 8 Jan 2022.

Rioual, Stephane, Benoit Lescop, Julien Pellé, Gerusa D. Radicchi, Gilles Chaumat, Marie D. Bruni, Johan Becker, and Dominique Thierry. 2021. Monitoring of the environmental corrosivity in museums by RFID sensors: Application to pollution emitted by archeological woods. *Sustainability* 13: 6158.

Schito, Eva, Paolo Conti, Luca Urbanucci, and Daniele Testi. 2020. Multi-objective optimization of HVAC control in museum environment for artwork preservation, visitors' thermal comfort and energy efficiency. *Building and Environment* 180: 107018.

Strlič, Matija, David Thickett, Joel Taylor, and May Cassar. 2013. Damage functions in heritage science. *Studies in Conservation* 58: 80–87.

Wickham, Hadley. 2014. Tidy data. *Journal of Statistical Software* 59: 1–23.

Wickham, Hadley, Mara Averick, Jennifer Bryan, Winston Chang, Lucy D. McGowan, Romain François, Garrett Grolemund, Alex Hayes, Lionel Henry, and Jim Hester. 2019. Welcome to the tidyverse. *Journal of Open Source Software* 4: 1686.

Wickham, Hadley, and Garrett Grolemund. 2016. *R for data science*. California, CA: O'Reilly Media Inc.

Zhang, Xincheng, Peilin Dai, and Ze Zhao. 2021. Reflections on the indoor environmental monitoring system of the heritage buildings in the Palace Museum—a case study of the Meridian Gate Exhibition Hall. The International Archives of the Photogrammetry, Remote Sensing and Spatial Information Sciences, Volume XLVI-M-1-2021. 28th CIPA Symposium "Great Learning & Digital Emotion", 28 August–1 September 2021, Beijing, China.

Museums Zenithal Lighting: Studies of the Skylight in Museo De l'Almoina in Valencia (Spain)

José-Luis Baró Zarzo, María-Antonia Serrano, Juan-Carlos Moreno Esteve, and Fernando-Juan García-Diego

Abstract This work makes a brief description of zenithal lighting used in museums, and other heritage sites. This type of lighting has interesting advantages: it produces a good level of illuminance with a correct color spectrum. In addition, it allows the observer to perceive the sensation of being outdoors, while protecting from wind and rain. However, skylights can also filter radiation that can be detrimental to the preservation of the assets below, as well as causing very marked gradients between the illuminance of the areas just below and those further away. Using l'Almoina Archaeological Museum (Valencia, Spain) as a case study, the aim of the present study consists of evaluating the transmittance of visible, ultraviolet and near infrared solar radiation through the large skylight that illuminates much of the interior space. This transmittance varies throughout the year and time of day. Empirical data were taken on summer and winter at three different times around noon using two spectrometers: one for the ultraviolet bands (A: 280–315 nm; B: 315–400 nm), and other for both visible (400–700 nm) and near infrared (700–1000 nm) bands. Several points under the skylight inside the museum were measured and quantified by the percentage of transmittance. A discussion in how this kind of illumination can affect the cultural heritage that houses, and the results of other zenithal lighting used in architecture is done.

J.-L. B. Zarzo
Department of Architectural Composition, Universitat Politècnica de València, Valencia, Spain
e-mail: jobazar@cpa.upv.es

M.-A. Serrano
Centre for Biomaterials and Tissue Engineering, Universitat Politècnica de València, Valencia, Spain
e-mail: mserranj@fis.upv.es

J.-C. M. Esteve · F.-J. García-Diego (✉)
Department Applied Physics, Universitat Politècnica de València, Valencia, Spain
e-mail: fjgarcid@upv.es

J.-C. M. Esteve
e-mail: jcmestev@fis.upv.es

Á. F. Perles-Ivars et al. (eds.), *Collection Care*, Springer Proceedings in Archaeology and Heritage, https://doi.org/10.1007/978-3-031-85655-6_6

1 Introduction

According to Boito (1893), a monument should be consolidated rather than repaired, and repaired rather than restored. Conservation and preventive maintenance constitute a preliminary step that can avoid having to reach more pronounced and perhaps irreversible stages of degradation. This is recognized in the main international charters on the protection of archaeological heritage (The Venice Charter 1964; ICOMOS Charter 1990).

Among the different ways to prevent degradation, the control of light radiation is especially relevant in the case of museums (Commission Internationale de L'éclairage 2004; Conservación del Patrimonio Cultural 2019). Effectively, excessive exposure to visible light can cause pigment discoloration; the infrared radiation causes warming of the surface of objects; and the ultraviolet light brings about yellowing and disintegration of some materials.

In fact, the over exposition of assets to solar radiation, enclosed in totally or partially glazed structures, has originated serious troubles in the Roman villa of Piazza Armerina, Sicily-Italy, the ruins of the medieval cathedral of Hamar, Norway, and the Roman baths of Badenweiler, Germany, among others.

Regarding this context, we will address the archaeological site of l'Almoina in Valencia, Spain, as a case study to analyze the conservation conditions of the reminds found from different historical epochs, the oldest of which dates back to the founding of the city in the Roman period. It contains partial foundations of the forum, the curia, the basilica, the baths, and the sanctuary of Asklepios (Ribera 2009, 2012).

In order to preserve the archaeological site, a museum was erected over the remains in 2006, containing a large skylight of 300 square meters (17 × 17 m). The skylight allows capturing intense natural light, producing at one time an effect of transparency that attracts passers-by' attention, and suggesting to visitors the effect of an open space (Fig. 1).

Regardless of the good reception by the public, the fact is that the presence of the skylight has been causing serious practical problems, such as the extensive maintenance required, the thermal oscillation, or the energy inefficiency in summer (García 2013a, 2013b; Moreno 2018).

Under a heritage point of view, the construction of this museum has meant an ambivalent criticism. On the one side, it has allowed the dissemination of the remains for the better understanding of the city history, and has generated a large urban space for people to enjoy, provided with quality modern architecture in close dialogue with pre-existing buildings. On the other side, it has led to a *svetramento* of the very heart of Valencia that has done nothing but consolidate the distortion produced in the medieval urban fabric after the demolition of the old Almoina and other annexes buildings, to which those that had already disappeared to enlarge the Plaza de la Reina and the Plaza de la Virgen must be added.

Literature provides different contributions on the preservation of archaeological sites exposed to solar radiation with specific approaches. They include texts from the most theoretical ones of Michalski (1987, 1997), and Çetin and İpekoğlu (2013),

Fig. 1 Exterior and interior of the Museo L'Almoina in Valencia, with the large glazed skylight

to the most technical of Camuffo (2014); Tuchinda et al. (2006), Al-Obaidi et al. (2014) or Horie (1980), among others.

The present study aims to evaluate the solar radiation coming through the skylight into the interior of the museum with regard the solar radiation received, and how it changes over time.

2 Materials and Methods

2.1 Materials

The skylight is made up of a laminated glass of 3 sheets of 10 mm thick separated by 2 sheets of butyral (PVB), on which a sheet of water extends to enrich the effects of projection and filtering of light. Butyral increases the bending strength of the glass, improving safety in case of breakage and reducing the transmittance of ultraviolet radiation. The skylight is located just above de Roman baths' remains.

As shown in Fig. 2, only a part of the irradiance incident on the surface of the skylight is transmitted to the interior, as much of it is absorbed or reflected as it passes through the different layers that make it up.

The magnitude that has been managed refers to the ratio between the irradiance captured on the outside and the irradiance transmitted to the inside for each wavelength. This magnitude has been called Relative Attenuation (RA).

Two different spectrometers with their capture terminals were used for data collection so that a broad spectrum of radiation was tested: FLAME-S-UV–VIS to measure de UV band (300–400 nm), and HR4000CG-UV-NIR to measure VIS (400–700 nm) and NIR (700–1000 nm), made by Ocean Insight. Both devices were calibrated in 2017 with a measurement uncertainty of approximately 10% across the entire measurement spectrum. In addition, a laptop computer was performed to record the data using two different applications by Ocean Optics.

2.2 Methods

For the development of the study several variables were introduced: location (points A, B, C); season (summer, on two different days, and winter, on only one day); time (at 9 a.m., noon and 3 p.m., always solar time, with three hours before and after midday); and presence of water on the glass (with water on July 30th, and without water: July 25th).

To capture the data clear, cloudless days were chosen. The terminal sensor was first directed outside towards the sun, and immediately the measurement was similarly recorded indoors, at 17 different points located both below and around the

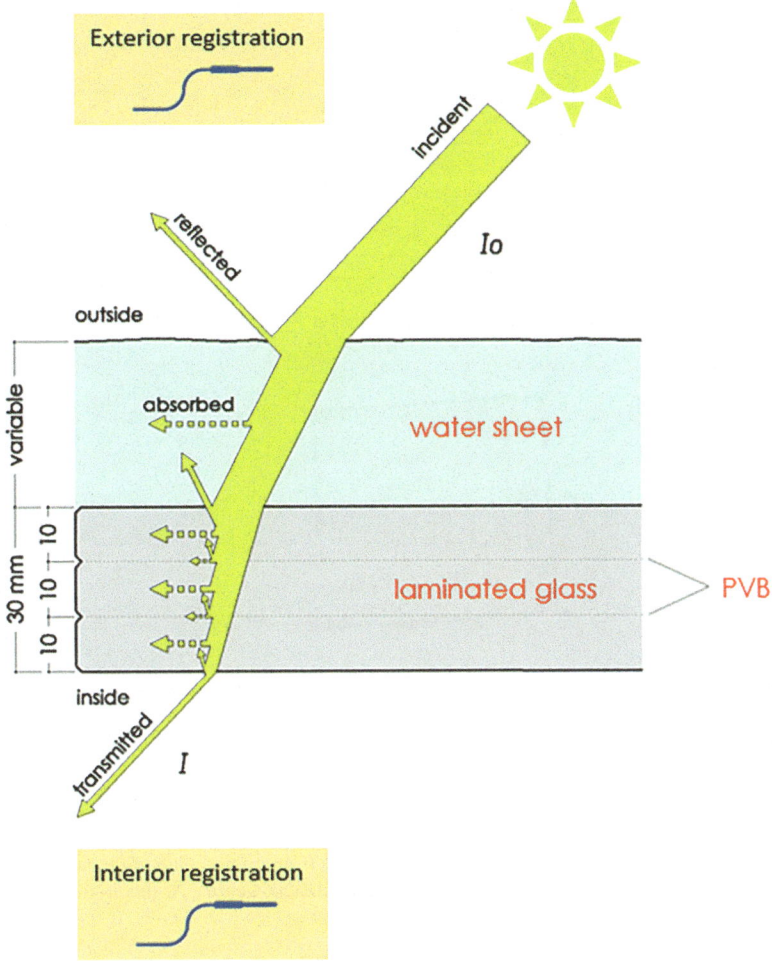

Fig. 2 Diagram of optical transmission through the skylight and water sheet. The incident intensity (I_0) is reduced progressively as a consequence of the partial absorption and reflection experienced when passing through the different layers, meaning that only a fraction is transmitted to the interior (I): $I \leq I_0$

skylight projection (Fig. 3). In all cases, two successive captures were taken to dismiss eventual errors.

For subsequent operational calculations, only the most representative points under the skylight were selected: point A, close to the West side; point B, close to the NW angle; and point C, in the center of the square defined by the skylight.

The aforementioned phenomenon whereby part of the incident light is reflected, part is absorbed, and part passes through when it reaches a semitransparent surface, can be expressed by a balance of irradiances for each wavelength as follows:

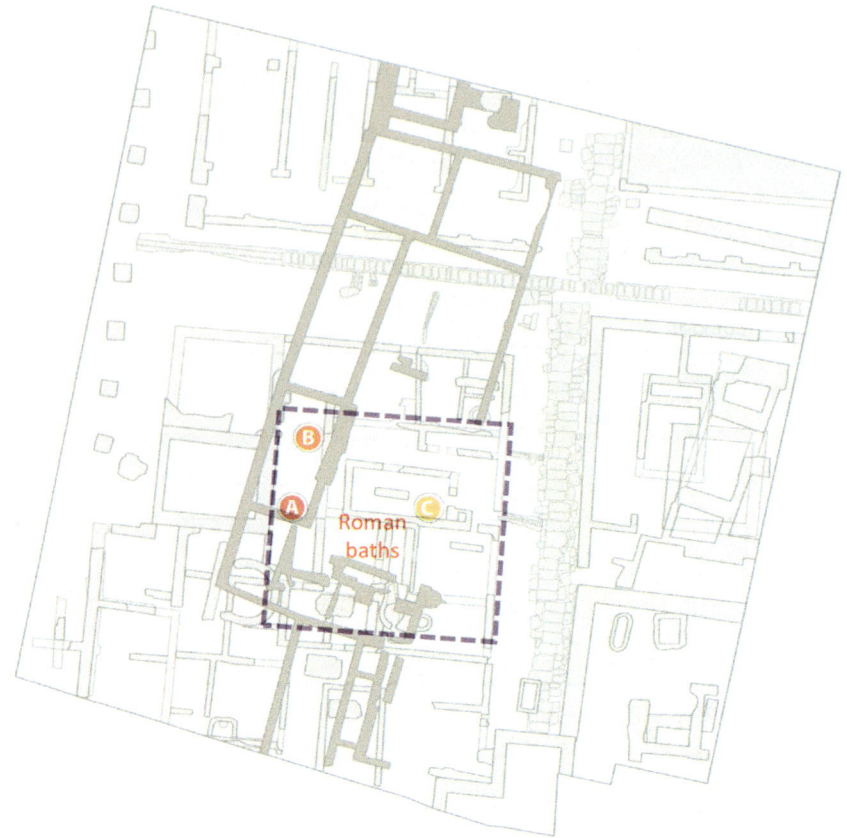

Fig. 3 Spatial location of the selected points. (Reworked from ground plan of the museum. Available online: https://www.esturismo.eu/Europa/Espana/Valencia/Centro_Arqueologico_de_la_Alm oina.html (accessed on 29 December 2021))

$$Irrad_{incident,\lambda} = Irrad_{reflected,\lambda} + Irrad_{absorbed,\lambda} + Irrad_{transmitted,\lambda} \qquad (1)$$

Optical relative attenuation (RA) defines a relative quantity (therefore dimensionless) that measures the fraction of incident light passing through a specific sample (Eq. 2), in our case, the glass of the skylight and eventually the water layer.

$$RA = I_{max}/I_{0,max} \qquad (2)$$

where I_{max} represents the transmitted light maximum intensity, and $I_{0,max}$ the incident maximum ray intensity, data which were provided by the two spectrometers cited above.

It also has been calculated an average value for each wavelength range studied in order to evaluate the comparative incidence of the different spectral bands (UVB,

UVA, VIS and NIR) for a given point (A, B, C), time (9 AM, noon, 3 PM) and day (July 25th and 30th, January 14th) according to the following expression (Eq. 3):

$$RA_{band} = \frac{\int I_\lambda dy}{\int I_{0\lambda} dx} \tag{3}$$

expression in which $I_{\lambda,max}$ represents the irradiance transmitted through the skylight, and $I_{0\lambda,max}$ the normal solar irradiance for each wavelength range. In this case, the step provided by the spectrometer (0.10 nm) was taken as the wavelength increment for calculation purposes.

Selected data were arranged in tables and presented in graphs using Microsoft Excel, then put in parallel for a clear visualization of the results.

3 Results and Discussion

We will distinguish among different approaches.

3.1 Measured Relative Attenuation Discussion

The relative attenuation concept we use does not correspond to the known definition of transmittance (International Lighting Vocabulary 2016). Transmittance is obtained experimentally under laboratory conditions by measuring the intensity of the incident and transmitted rays with directional devices. The measurement performed in this study is the minimum attenuation at a specific point, since it represents the ratio between the maximum incident intensity captured outside and the maximum registered inside at a lower point on the same day and time.

The methodology proposed here may be useful for application in museums with natural lighting, since the maximum intensity at a given point depends on many architectural factors (size of the skylight, position in the museum, orientation, shadows cast by other nearby buildings...), and therefore the only way to know it is by measuring it empirically.

According to the recommended standard for a stone museum (PD CEN/TS 16,163:2014 2014), lighting is not a factor affecting preventive conservation. However, it must be considered that, in this space, there are showcases containing valuable objects, and that it was also designed to host temporary exhibitions.

3.2   Spatial Study of the Spectral Relative Attenuation (UV and VIS–NIR Bands)

The RA of the UV band was calculated according to the Eq. (2) for the three locations, the three hours, and the two seasons chosen, as mentioned. The data collected from location C are represented graphically in Fig. 4 (top). It shows that RA is maximum (30%–100%) at 300 nm (on the left area), minimum (0%–5%) between 320–370 nm (on the central area) and rises reaching 10%–40% at 400 nm (on the right area). Similar tendencies have been observed in all locations, times, and seasons. Nevertheless, RA is slightly higher in winter than in summer.

As regards the VIS and NIR bands, the same procedure was followed. However, the results were widely dispersed in all variables, as shown in Fig. 4 (below). The highest RA at point C occurred in the morning, reaching 70% at 520 nm in summer, and 50% at 1000 nm in winter. NIR relative attenuation at location A is significant in winter in the morning and at noon, reaching 70% at 1000 nm. Meanwhile, in location B there is greater RA at 520 nm, reaching a value of 50% in summer at noon and 38% in the afternoon.

3.3   Mean Relative Attenuation Distributed Per Bands and Times

The mean relative attenuation (mRA) of the UV-B and UV-A bands was calculated using Eq. (3). The graphical representation of summer for the three locations is shown in Fig. 5 (top).

Substantial differences between summer and winter on the one hand, and between UV and VIS–NIR bands on the other appeared. In summer, the mRA of the VIS band is higher than the one of the NIR band. In winter, on the contrary, the mRA is highest at 12 h, and no significant at 15 h in both bands.

In the UV-B band, the highest mRA (20%) is reached in locations B and C in the morning of both seasons, while at midday and in the afternoon the mRA remains fairly constant throughout the museum (5–10%). Meanwhile, the mRA of de UV-A range is also quite constant ($\leq 3\%$), but lower than that of the UV-B band.

Regarding the VIS and NIR bands, in summer the highest mRA occurred in points A and B, as shown in Fig. 5 (bottom), and decreased from morning to afternoon, but with lower values in the NIR range (morning 35%, afternoon 15%) than in the VIS one (morning 50–65%, afternoon 30%).

In winter, within the same two ranges, the highest mRA reached the maximum value (85%) at noon in location A. In the morning, it was higher in the NIR range than the VIS range (30% vs. 10%), presenting little variation throughout the entire museum. In the afternoon, in both ranges, the mRA was less than 1% at all of the studied locations.

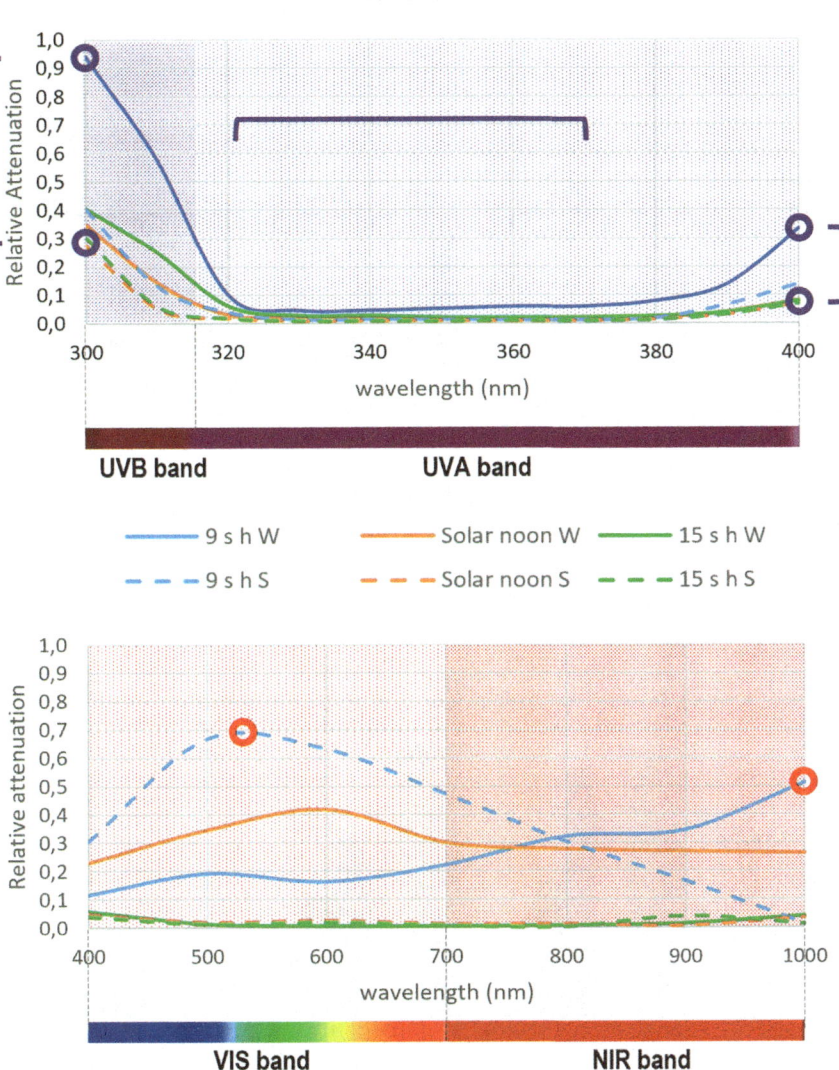

Fig. 4 Representation of the spectral relative attenuation for the UV bands (top) and VIS–NIR bands (bottom) for location C

3.4 Water Sheet Influence

With respect to the influence of the water sheet, it can be said that trends were similar in all variables. In general, there were little differences in RA considering the influence of the water sheet or not. Slight differences appear in the UV-B band

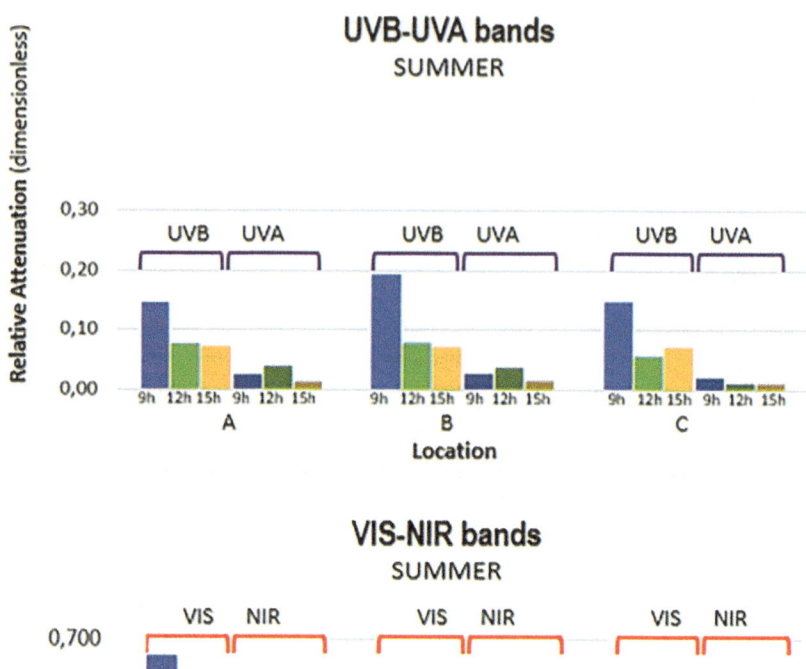

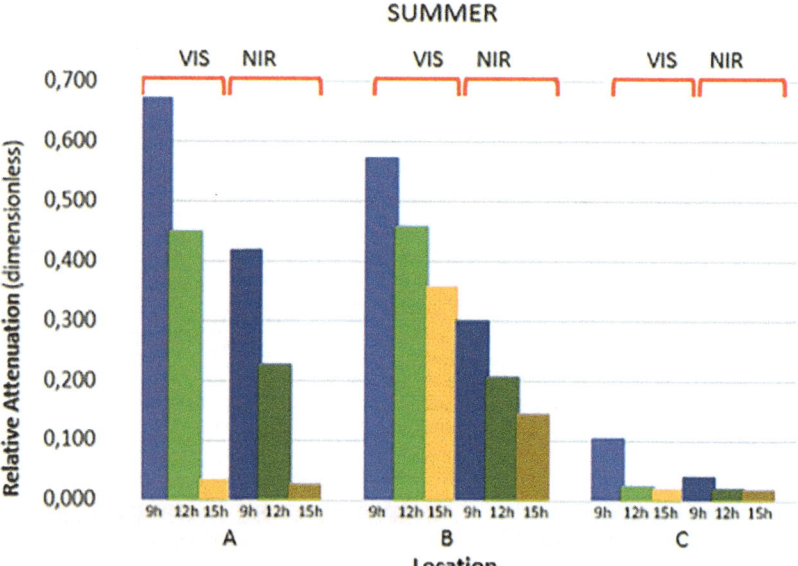

Fig. 5 Representation of the mean relative attenuation for the UV ranges (top) and VIS–NIR bands (bottom), for the 3 locations studied, in summer, at the three different times

and between 380 and 400 nm in UV-A. In most cases, RA is higher without water, which means that water contributes to the reduction of transmittance.

4   Conclusions

Results can be summarized as follows. For the VIS band the maximum RA was 70% reached at 525 nm (in cyan) in summer at locations A and C. In the case of the N-IR band, the maximum RA reached the 80% in winter. In the UV range the skylight RA is higher in the UV-B band (20%) than in the UV-A band (around the 5%). The observed results of a lower RA at higher temperature, as occurs at midday (in orange), could also be due to the phenomenon that as temperature increases, thermal conductivity decreases, leading to higher thermal inertia. The same explanation involves the fact that in summer RA were higher at 9 h solar than at solar noon and were similar to those in winter.

In general, measurement of RA has been found to be a useful tool in the preventive conservation of cultural heritage since the variability of values is conditioned by the entire specific architectural and decorative structure of the analyses space. Attention to the UV-B band around 300 nm should be paid. Another important conclusion is the demonstration that the water layer on the skylight increases the relative attenuation of the solar rays that fell upon it. Nevertheless, the actual data collection conditions give inconsistent values in some specific areas due to complex factors, such as the shadows cast by adjacent buildings or indeed by the Museum structure itself.

Among the avenues open to research, it would be worth mentioning the in-depth study of the particularities of water as a filter through laboratory essays under more objective working conditions, considering different thicknesses and additives. Additional research could address the possibilities of the structure itself to improve its behaviour in order to reduce the UV-B transmittance. Finally, a study of chromatic quality could be approached from the point of view of the correct perception of exposure by the observer, taking into account the characteristics of the transmittance of the visible spectrum wavelength band.

Acknowledgements The authors would like to thank the Ajuntament de València and Vicent Escrivà Torres, director of the Museo l'Almoina, for their assistance and authorization of the data collection without which this work could not have been carried out.

Funding This project received funding from the European Union's Horizon 2020 Research and Innovation Programme under Grant Agreement No. 814624.

References

Al-Obaidi, Karam, Mazran Ismail, and Abdul Rahman. 2014. A review of skylight glazing materials in architectural designs for a better indoor environment. *Modern Applied Science* 8: 68–82.
Boito, Camillo. 1893. Voto conclusivo del 3° congresso degli ingegneri e architetti Italiani tenutosi a Roma nel 1883. In *Questioni pratiche di belle arti, restauri, concorsi, legislazione, professione, insegnamento*, 28ff. Milano: Hoepli.
Camuffo, Dario. 2014. Radiation and light. Conservation, restoration, and maintenance of indoor and outdoor monuments. In *Microclimate for cultural heritage*, 131–164. Amsterdam: Elsevier.

Çetin, Funda Y., and Başak İpekoğlu. 2013. Impact of transparency in the design of protective structures for conservation of archaeological remains. *Journal of Cultural Heritage* 14: e21–e24.
Commission Internationale de L'Éclairage. 2004. *Control of damage to museum objects by optical radiation.* Vienna: Fer.
Conservación del Patrimonio Cultural. 2019. *Especificaciones para el emplazamiento, construcción y modificación de edificios o salas destinadas al almacenamiento o utilización de colecciones del patrimonio; UNE-EN 16893:2019.* Madrid: Asociación Española de Normalización (UNE).
García, H. 2013a. El estanque de la plaza de la Almoina se sustituirá por un lucernario piramidal. *Levante-EMV*, May 19.
García, H. 2013b. Descartan la pirámide de cristal de la Almoina por el efecto sauna en el museo. *Levante-EMV*, June 5.
Horie, Charles V. 1980. Solar control films for reducing light levels in buildings with daylight. *Studies in Conservation* 25: 49–54.
ICOMOS Charter. 1990. Charter for the protection and management of the archaeological heritage. http://wp.icahm.icomos.org/wp-content/uploads/2017/01/1990-Lausanne-Charter-for-Protection-and-Management-of-Archaeological-Heritage.pdf. Accessed 4 July 2021.
International Lighting Vocabulary. 2016. International commission on illumination CIE DIS 017/E:2016 ILV. https://cie.co.at/e-ilv. Accessed 24 Oct 2021.
Michalski, Stefan. 1987. Damage to museum objects by visible radiation (light) and ultraviolet radiation (UV). In *Lighting in museums, galleries and historic houses. Papers of the conference,* 3–16. Bristol.
Michalski, Stefan W. 1997. The lighting decision. In *Fabric of an exhibition, preprints of textile symposium 97,* 97–104. Ottawa: Canadian Conservation Institute.
Moreno, Paco. 2018. El centro arqueológico de la Almoina de Valencia se reformará a los once años de su apertura. Las Provincias. https://www.lasprovincias.es/valencia-ciudad/ayuntamiento-encarga-estudio-almoina-20181024131618-nt.html. Accessed 4 Jan 2022.
PD CEN/TS 16163:2014. 2014. Conservation of cultural heritage. Guidelines and procedures for choosing appropriate lighting for indoor exhibitions. https://standards.cen.eu/dyn/www/f?p=204:110:0::::FSP_PROJECT:34047&cs=1AFCAEA358660F36ECF4D92D51A8AD2FC. Accessed 4 Nov 2021.
Ribera, Lacomba A. 2009. El centro arqueológico de L'Almoina. Valencia. In *Proceedings of the 5 encuentro internacional actualidad en museografía,* 67–82. Palencia, Spain.
Ribera, Lacomba A. 2012. El centro arqueológico de L'Almoina en Valencia. In *Archeologia e città: Riflessione sulla varizzazione dei siti archeologici in aree urbane,* Palombi ed. 37–45. Roma: Ministero dei Beni e delle Attività Culturali e del Turismo Soprintendenza Speciale per i Beni Archeologici di Roma.
The Venice Charter. 1964. Committee for drafting the international charter for the conservation and restoration of monuments. International charter for the conservation and restoration of monuments and sites. https://www.icomos.org/charters/venice_e.pdf. Accessed 4 July 2021.
Tuchinda, Chanisada, Sabong Srivannaboon, and Henry W. Lim. 2006. Photoprotection by window glass, automobile glass, and sunglasses. *Journal of the American Academy of Dermatology* 54: 845–854.

Sustainability in Museum Lighting

David Saunders⑩

Abstract The environmental sustainability of museum lighting is placed within the context of wider strategies to reduce the use of non-renewable energy. Various measures to reduce energy and resource consumption are presented, including switching to, or generating, renewable electricity, the installation of more energy efficient and longer lasting LED lamps, increased use of daylight and changes to museum lighting practice that reduce energy consumption. The sustainability of lamps and lighting systems is explored in the respect to the precepts of reduction, reuse and recycling of materials. A method of considering environmental sustainability alongside economic, operational, social and societal sustainability is described. The model is used to assess the impact of the various changes to museum lighting practice introduced to improve environmental performance on these other measures of sustainability, paying particular attention to their effect on social sustainability (defined in terms of current accessibility) and societal sustainability, which encompasses the longer-term preservation of objects.

1 Cultural Heritage Organisations and Environmental Sustainability

The drive to achieve net zero energy continues to make international headlines, with recent coverage centred on the COP26 meeting in Glasgow. While the public sector may consume a small proportion of national energy use, and cultural heritage organisations a small proportion of that amount, every sector needs to contribute to these targets. The UK government's Carbon Plan (HM Government 2011) states that "While the public sector represents only around 3% of the UK's greenhouse gas emissions, it has a responsibility to lead the way in reducing them" and there is widespread acknowledgement that museums must play their part if global targets for climate change reduction are to be met, exemplified by the recent *Joint Commitment*

D. Saunders (✉)
Department of Scientific Research, The British Museum, Great Russell St, London WC1B 3DG, UK
e-mail: david@saunders-online.net

Á. F. Perles-Ivars et al. (eds.), *Collection Care*, Springer Proceedings in Archaeology and Heritage, https://doi.org/10.1007/978-3-031-85655-6_7

71

for Climate Action in Cultural Heritage by three leading international conservation bodies, IIC, ICCROM, and ICOM-CC (IIC et al. 2021).

Data on museum energy use are not particularly easy to find, as they are generally agglomerated into broader categories. In the USA *public assembly buildings* consume around 0.8% of total national energy use, but there are eight classes of such building including those for *entertainment or culture*, of which museums represent one of six further sub-categories (USEIA 2012). In the UK, museums consume around 3% of "community, arts and leisure" use, itself a small proportion of public sector consumption (BEIS 2016). Data on the breakdown of energy consumption within museums or other public sector buildings indicate that most of this energy (*c.*75%) is consumed in heating, cooling, ventilation, and humidification (BEIS 2016; USEIA 2012). To reduce energy consumption and greenhouse gas emissions by museums, great emphasis has often been placed on balancing the need for appropriate climatic conditions for collections and visitors against energy use.

One route to reduce emissions has been to develop more energy efficient buildings and systems, favouring passive or low technology methods over mechanical control of the internal climate. This has proved particularly effective for storage facilities, where low occupancy and simpler architecture often help to reduce the need for active control. The conservation guidelines for temperature and relative humidity, and variations in these parameters, have also been reassessed to determine whether broader tolerances might permit energy savings to be made without threatening the long-term preservation of collections. The greater differentiation between objects that can or cannot withstand wider fluctuations in their environment allows resources to be targeted to the most vulnerable parts of a collection, by providing tighter environmental control in certain rooms, display cases or microclimate enclosures.

2 Lighting Within the Environmental Sustainability Discussion

Because it represents a lesser proportion of energy usage—typically around one sixth (BEIS 2016)—lighting has often been a secondary consideration in discussions of environmental sustainability. However, when considering electrical energy usage within museums, lighting constitutes a greater proportion of consumption, often up to one third (BEIS 2016; USEIA 2003). Although lighting can be a major contributor to electrical energy use, a study of public sector buildings in the UK in the late 2010s estimated that changes to lighting equipment and practice could constitute around a third of the potential electrical energy savings in non-residential buildings by 2032 (BEIS 2018).

Reducing the effect of lighting on a museum's carbon footprint can be approached from four directions. First by considering the emissions inherent in the generation of the electrical energy used to power the lighting systems, secondly by improving the energy efficiency of the lamps and other lighting equipment, thirdly by greater use of daylight when possible and, finally, by improving practice to avoid unnecessary light/energy use. In addition to these measures to reduce greenhouse gas emissions, another significant environmental sustainability criterion is that of resource use and reuse. As in other spheres, resource sustainability in lighting relies on reducing use, reusing where possible and recycling materials.

3 The Greenness of Electrical Energy

Even in museums that employ a significant proportion of available daylight, the degree to which lighting contributes to greenhouse gas emissions depends strongly on the source of the electrical energy. A completely carbon–neutral option is, of course, for the institution to generate its own electricity, generally using photovoltaic (solar) panels, although other options such as geothermal and hydroelectric generation are possible, depending on location. The extent to which museums can rely solely on solar energy will depend on electricity use after other energy saving measures have been implemented and the local availability and strength of daylight.

For buildings—particularly stores—with a relatively low energy consumption, rooftop installations may provide sufficient electricity to meet annual needs, for example the Ornithology Building of the Natural History Museum at Tring, UK, where an array of 318 solar panels generates around 75 kWh per annum, reducing CO_2 emissions by 21 tonnes and simultaneously saving costs (NHM 2020). In other cases, space constraints make it difficult to install sufficient photovoltaic panels to serve an entire institution's needs unless there are grounds or buildings on which to site photovoltaic panels; in the USA, the Shelburne Museum has become self-sufficient in electricity by installing panels in an adjacent field, while the Toledo Museum of Art generates a high proportion of the institution's electricity from solar panels on roofs of the museum and visitor parking lot (Bintz and Bernard 2014). In practice, most institutions remain connected to the electricity grid and may be net consumers or producers of electricity at different times of the day or year.

Museums that rely more heavily on electricity from the grid can opt to receive 'green' energy, generated from renewable sources, although it needs to be understood that there is seldom a direct relationship between a green source of energy and a green energy supplier. There are also many reports of 'greenwashing', ranging from the supply of electricity certificated as green (for example, using *Renewable Energy Guarantees of Origin*—REGOs) but produced from fossil fuels, to application of the term 'green' to an electricity supply of which only a small proportion is generated from renewable sources. Nevertheless, end user demand for electricity generated from renewable sources plays a role in increasing demand for the move away from fossil fuels. In the decade to 2020 the renewable energy share of global electricity capacity has risen from around 25 to 37% (IRENA 2021:48). In some countries, particularly those that have established hydroelectric, solar or geothermal energy sectors, the figure approaches 100% (IRENA 2021:48–50).

4   The Efficacy of Electric Lighting

An area over which museums have rather more control is the selection of electric lighting systems that make more efficient use of energy. In the USA lamps that use 90% less energy than traditional incandescent lamps bear the Energy Star® symbol (Energy Star 2021). Lamps can also be classified according to their efficacy, that is the visible light output per unit of electrical energy consumed—measured in lumens per Watt (lumen.W^{-1}); this measure is used by the European Community (EC) to assign lamps to particular categories of energy efficiency.

Table 1 gives the range of efficacies for lamp types commonly found in museums in the late twentieth and twenty-first centuries along with their designations under the 2012 and 2019 EC energy rating systems (EC 2012a, 2019a). It can be seen that tungsten incandescent lamps, which were the mainstay of museum lighting throughout the first half of the twenty century, are relatively inefficient, leading to their gradual replacement by rather more energy efficient (tungsten) halogen lamps and much more efficacious fluorescent sources over the remainder of the century. The introduction of LED lamps has already greatly increased the potential light output per Watt, to the extent that the highest rating in the 2012 EC system (A) had been subdivided to include A + , A + + and A + + + categorisations as lamp efficacy improved. The new 2019 EC system, which came into force in September 2021, resets the boundaries for each rating, leaving space for potential future improvements in lamp technology. Although there are a number of LEDs with advertised efficacies of 200 lm.W^{-1}, there are currently no A rated commercial lamps, the threshold for which is 210 lm.W^{-1} (EC 2019a).

The theoretical maximum lamp efficacy is 683 lm.W^{-1}, but this corresponds to a monochromatic green light source. While it has been calculated that a white source with using red, green and blue LEDs can achieve an efficacy of around 400 lm.W^{-1}, this assumes 100% quantum efficiency, leading to the suggestion that the maximum efficacy achievable from a white lamp is likely to be around 250–300 lm.W^{-1} (Ohno 2006). Some of the advances in LED technology that might lead to greater efficacy

Table 1 Luminous efficacy, energy ratings and average rated lamp life for common electric light sources

Description	Luminous efficacy (lumen.W^{-1})	2012 EC energy rating	2019 EC energy rating	Average rated lamp life (hours)
Tungsten incandescent lamp	5–15	E or worse	G	1000–1500
Halogen lamp	15–25	B–C	G	2000–4000
Compact fluorescent lamp	35–80	A–B	G–F	6000–10,000
Linear fluorescent lamp	70–100	A	G–F	10,000–20,000
White LED	50–200	A–A + + +	G–B	25,000–50,000

include: improvements in the efficacy of the semiconductor devices—particularly in the growth of the n- or p-type semiconductors; new or improved phosphor materials with higher quantum efficiencies; tailoring the diode emission to maximise phosphor excitation; and better optics (encapsulation) to increase the efficiency of light delivery from the diode.

In the immediate future at least, LEDs seem set to be the dominant lighting technology, a position reinforced by the progressive banning of other types of lamps in many countries. In Europe, for example, incandescent lamps were banned in the 2010s, certain types of fluorescent and halogen lamps on 1 September 2021 and, with some exemptions for specialised applications, most remaining fluorescent and halogen lamps will be banned after 1 September 2023 (EC 2019b).

Another driver for the replacement of existing lamps by LEDs is the latter's much longer lifetime. Intercomparison of lifetime and deterioration in output over time is not straightforward, but Table 1 gives an indication of the relative lifetimes of different lamp technologies, demonstrating the longer period between relamping offered by LEDs. Less frequent lamp changes can contribute to reducing waste and have important implications for resource sustainability—see the section on resource use and reuse.

Alongside the environmental benefits of lower energy consumption and reduced waste, lower running costs have frequently proved an important driver in the switch to LED lighting. For example, the replacement of halogen lamps with LED sources across the National Gallery, London, was calculated to offer annual energy savings of 85% (over GB£ 50 k) in addition to a reduction in CO_2 emissions of over 400 tonnes per annum (Padfield et al. 2013). Longer term savings in energy bills and through less frequent relamping are partially offset by the installation cost of new systems, with payback period often cited alongside reductions in running costs and CO_2 emissions when assessing the success of new LED installations. In the Fossils Gallery at the Manchester Museum, UK, the installation of LED lamps resulted in an 89% reduction in energy consumption and emissions (equivalent to 60 tonnes of CO_2 per annum), which was sufficient to pay back the investment in the lighting system after 18 months (MLA and ARUP 2010). A study commissioned by the US Department of Energy of the installation of LEDs in the Brooker Gallery at the Field Museum, Chicago determined a payback period of *c*.39 months based on the electrical energy consumed by the lighting system. However, the higher efficacy of the LED lamps led to lower heat generation, a reduced load on the cooling system and consequent energy savings that shortened the payback period to *c*.28 months (Myer and Kinzey 2010).

5   Increasing Daylight Use

Until the introduction of more efficient gas lighting in the late nineteenth century, collections were seen predominantly under daylight and, consequently, only at those times of day when sufficient light for viewing was available. The advent of gas and electric lighting has allowed comfortable viewing conditions to be maintained during a greater portion of the year in regions away from the equator where daylight availability alters seasonally, in some case to the extent that galleries are lit wholly by electric light.

While a return to a greater reliance on daylight is attractive from an environmental sustainability perspective, the reduction in energy consumption, resource use in lamp manufacture and associated greenhouse emissions is offset by three principal drawbacks. First, the proportion of infrared (IR) and damaging ultraviolet (UV) radiation in daylight is appreciably higher than that of most electric light sources, particularly the LED lamps that play an increasing role in museum lighting. The issue of UV can be addressed relatively easily by ensuring that daylight is only admitted to display areas through glazing that absorbs the majority of UV radiation. The potential heat gain from IR radiation (solar gain) can be mitigated by using low emissivity (low-E) glazing and by ensuring that any shading—see following paragraph—is applied externally.

Second, the availability of daylight is highly variable; seasonally, diurnally and geographically. These factors will dictate the extent to which daylight will need to be supplemented by electric light if a comfortable viewing environment and acceptable levels for preservation are to be provided while maintaining regular opening hours. Daylight will also be subject to short term changes in availability, direction and colour during the course of a day or as weather conditions change, which make it difficult to ensure good viewing conditions for a time-limited visit to the museum.

Arising from this variability, a third drawback of using daylight use is that it may require a higher degree of control if conditions for viewing and preservation are to be balanced. Over-control of daylight has been eschewed in the past, largely because its variability in direction, colour and strength were valued and set it apart from electric lighting. If daylight is chosen for reasons of sustainability rather than aesthetics, greater control might be acceptable.

Any increased complexity and cost resulting from greater use of daylight will depend strongly on the museum building and the solutions available for daylight control. In newly built museums there is more scope for applying architectural solutions to the control of daylight, through careful consideration—often supported by computer modelling—of the number and orientation of skylights and windows, or the positioning of fixed louvres, baffles or frits to reduce overall or directional daylight ingress. It may also be possible to retrofit these passive solutions to existing buildings, although care may be needed to respect their historical appearance and structure. When introducing or re-opening windows or skylights some care is needed to ensure that sources of glare and veiling reflections are not created. Particularly for new interior spaces that might normally be lit with electric lamps, it is possible—although not entirely straightforward—to use light transportation systems such as light pipes or fibre optic cables to bring daylight into these spaces, as is the case in the lower floor rooms at the National Gallery of Canada.

Moveable blinds, louvres or curtains offer more dynamic control of changing daylight. Manually controlled systems require greater intervention but are less energy intensive and do not rely on complex sensor and control technologies. Manual shading is often linked to light plans or instructions that recommend set positions for particular times of day and seasons (Thickett 2016). Increasingly, rather than relying on periodic 'spot' measurements, this information on daylight distribution in exhibition spaces across days and seasons can be derived from imaging methods (Blades et al. 2020; Cannon-Brookes et al. 2017), or theoretical modelling based on pre-existing 3D data or relatively simple 3D scans of spaces and objects.

Better response to changing daylight availability may be possible using automated blind or louvre systems controlled by light sensors in the exhibition spaces, although care is needed to ensure that the constant adjustment of the blinds is not distracting, for example when clouds pass across the sun. Another, if as yet infrequently applied, approach is to install electrochromic glass linked to a light monitoring system that can increase the degree of shading in response to high daylight levels (Vlachou-Mogire et al. 2020). The extent to which such automated methods decrease reliance on electric light, and thus reduce energy consumption, needs so be balanced against and the energy used by the monitoring and control systems, the embodied carbon in their production (and periodic replacement).

6   Improving Practice

Some relatively simple changes to lighting practice can produce significant savings in energy use. Although the brief examples given here relate to lighting collections, many are equally applicable to other spaces within the museum and may already have been considered or instigated as part of wider energy reviews or policies. Most of these changes in practice take as a starting point the premise that lights should be off—or operating at a reduced level—when they are not required. In this respect they will be familiar to conservators acting to reduce damage to collections by minimizing light exposure. The extent to which lighting can be reduced during museum opening hours depends strongly on visitor numbers. Well-attended museums or displays will probably need to be lit constantly, but the lights in less frequented areas or less popular display cases may be raised in response to the presence of visitors. At their simplest, display case lights can be controlled by a visitor-operated switch, with more complex case and room systems sensing the presence of visitors through their motion or emitted infrared radiation. Zoning is a potentially efficient method of ensuring that only the required part of a space or collection is lit. While it can work well for offices, stores or study rooms, and is useful for switching lights in individual display cases, it can produce a peculiar appearance if applied to regions of a larger gallery.

While, extinguishing all lighting outside visiting hours might improve energy consumption and enhance preservation, some low-level lighting is likely to be needed for activities such as cleaning and security. Whether the systems for out-of-hours may be either manual or programmable, it is worth making checks to ensure that the lights are switched off, particularly after events that cause a departure from normal routine.

7 Resource Use and Reuse

There are a number of different models for resource sustainability, including the comprehensive nine principles of the circular economy defined by European Commission (Hirsch and Schempp 2020), but all include commitments to reduce, reuse and recycle materials (Fig. 1).

7.1 Reduce

For lamps and lighting systems this implies both the selection of equipment with a longer lifetime and a commitment to their replacement when they fail or become unserviceable rather than once they are no longer new or fashionable. From Table 1 it is clear that LED lamps offer longer lamp lives than the technologies used in the twentieth century, offering the possibility of less frequent replacement, but there is rather less information on the amount of embodied carbon involved in the manufacture of LED lamps or fittings, making it difficult to calculate the impact of the full life cycle of production and disposal of lighting systems. In a drive to extend further their lifetime, there is considerable research into the factors that cause LED lamps to fail, as the LED itself fails in around 10% of lamps and the failure of the associated electronics accounts for more than half of failures (US Department of Energy 2013). There is a common practice in museums to 'bulk relamp' rooms once a few of the lamps have failed. While this might make sense for operational and economic reasons, these need to be balanced against the environmental implications of the potential disposal or recycling of perfectly good lamps.

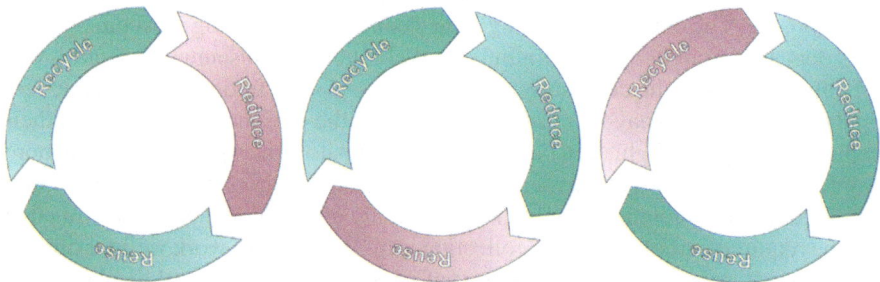

Fig. 1 Three tenets of improved resource management: Reduce, reuse and recycle

7.2 Reuse

Where there are compelling operational, economic or aesthetic reasons to replace lamps or systems, reuse within the institution, or resale or gifting to another organization, can provide an alternative to disposal or recycling. There have been huge advances in the sustainability of temporary displays or exhibitions with greater focus on reusing not just lighting systems, but other materials, particularly those used in display cases. The same policy of reuse is not always applied to lamps, but if there are reasons that these cannot be reused in the next exhibition, or are removed as part of a bulk relamp, it might be possible to employ them elsewhere in the museum, perhaps in less critical applications.

An emerging trend is that of leased lighting systems, in which the lamps and luminaires are typically leased rather than purchased. The contract may include a maintenance aspect and at the end of the lease agreement the lamps and luminaires are reused or recycled if the client wants an updated system. To date such leased systems have largely been limited to heterogenous lighting environments such as warehouses and offices, rather than the more complex lighting of museums and historic houses. Against their advantages, there is a danger that—in common with leased vehicles and mobile phones—such agreements may encourage systems to updated more frequently than is necessary. In addition, there is the question of whether such practices may promote the 'dumping' of used systems in less affluent societies.

7.3 Recycle

The boundary between reuse and recycling is blurred as, increasingly, schemes are developing to repair, refurbish, recondition or remanufacture products, including lighting systems. Definitions vary, but the relevant British Standards helpfully defines reconditioning as returning a product to a satisfactory working condition while remanufacturing provides a product with equivalent or better performance than the original (BSI 2010, 2011).

When fittings reach the end of their useful life or lamps fail recycling helps to maximize the use to which their component materials can be put. In Europe legislations covers the recycling of lamps and fittings within the directive on waste electrical and electronic equipment—WEEE (EC 2012b). There has been considerable focus on the safe disposal of lamps that contain small amounts of mercury, which include all fluorescent and most high intensity discharge lamps. These lamps are exempt from legislation than bans the use of mercury and other substances hazardous to health, including lead, cadmium, chromium, polybrominated organics used as fire-retardants and some phthalates used as plasticisers (EC 2011). Although fluorescent lamps continue to be phased out in favour of LEDs on energy efficiency grounds, their disposal will continue to be an issue for some time and schemes including the

European 'Eucolight' and US 'Lamp Recycle' initiatives promote their responsible disposal and recycling.

More broadly these and other programmes promote the recycling of all types of lamps and fittings. In 2019 the EucoLight scheme collected 52,000 tonnes of lighting products from 127,000 collection points (EucoLight 2019). Better methods of sorting waste lamps and fittings to recycle metals, glass and plastics—and for the recovery of rare-earth elements from fluorescent lamps—are under development. With the increased use of LEDs there has been a research focus not only on the recovery of indium, gallium, rare-earths and of other metals from these lamps, but on better initial lamp design to facilitate end-of-life recycling (Casamayor et al. 2018; Rahman et al. 2021).

8   Introducing Broader Sustainability Criteria

Until relatively recently the term 'sustainability' has been largely synonymous with the concept of conscious energy and resource management with the goal of reducing greenhouse gas emissions, pollution and the depletion of non-renewable reserves. Some consideration was often given to financial sustainability, particularly in calculating the cost benefits of measures to improve environmental sustainability. In addition, there has been an increasing recognition that people-centred criteria for assessing sustainability are an important balance to more easily quantified environmental and fiscal measures.

The current use of the term sustainability is, therefore, both broader and more nuanced, with its application not only to environmental issues but also to aspects of social justice and quality of life, best exemplified in the 17 Sustainable Development Goals (SDGs) defined by the United Nations (https://sdgs.un.org/goals). While the SDGs contain several goals focused on the environment and others relate to economic goals, over half relate to people-centred measures of sustainability such as equality and physical and mental wellbeing.

While, at an organisational level, we may well wish to embrace all these goals— or those that have relevance for our ambit—it may be more practical and pragmatic to develop a subset of these sustainability criteria in order to deal with particular decision-making processes within a museum or other cultural heritage organisation, for examples those concerning the lighting of objects and collections. A model based on five 'pillars' of sustainability is shown schematically in Fig. 2; its development is described briefly here and given in more detail elsewhere (Saunders 2022). The model introduces two pillars (social and societal) that embody the anthropocentric elements of the SDGs to balance against the quantitative considerations of energy use and cost embodied in the environmental and economic pillars common to many schemata. Broadly, the social pillar encompasses sustainability as it relates to individuals and small communities, or perhaps in the shorter term, while the societal pillar covers larger societies or cultures, and longer-term sustainability.

Environmental	Energy consumption; resource use; waste production; CO_2 emissions
Social	Display visibility; availability; enjoyment; comfort, pleasantness or ambiance
Societal	Collection lifetime; engagement (buy-in); cultural continuity
Operational	Complexity of control systems; complexity of practice
Economic	Energy savings; installation costs; maintenance and replacement costs

Fig. 2 The five 'pillars' used to model sustainable decision making and examples of criteria associated with each pillar [this figure must be reproduced in colour]

Importantly, when we come to make decisions about the display and preservation of collections, these pillars can also be used to map aspects of current and future access onto the concepts of social and societal sustainability, as indicated in Fig. 2. Current access can be seen as a way to promote enjoyment, engagement and the wellbeing of individuals and communities, the pillar of social sustainability. Future access—a derived benefit of preservation—adds to these social gains the benefits to cultures of continued access to collections that reflect and embody their creativity, traditions, history and continuity; the pillar of societal sustainability (Saunders 2022).

A fifth pillar, termed operational, captures the practicalities—particularly in the longer term—of systems and practices that might be introduced to meet other sustainability goals, which are often neglected when initial decisions are reached. A definition of the criteria associated with each of the pillars will probably need to be developed at an institutional level, but Fig. 2 gives examples to illustrate how the factors that affect decision making for presentation, storage and preservation might be mapped in terms of these considerations of sustainability.

9 Applying the Sustainability Model to Museum Lighting

Having now developed a conceptual model based on five pillars and defined how aspects of preservation and access map onto these sustainability criteria as people-centred outcomes, we can apply it to a series of examples, seeing how some of the strategies described above that aim to improve the environmental sustainability of museum lighting affect other measures of sustainability.

9.1 *Example 1: Buying or Generating 'Green' Energy*

This case study is explored in greater detail in the paper describing the development of the sustainability model (Saunders 2022). Example 1A in Table 2 illustrates that a simple switch of electricity supplier or tariff is unlikely to affect operational sustainability or, in the absence of other changes to practice, either of the two anthropocentric indicators of sustainability. With the caveats described above concerning the legitimacy of the renewable origin of the supply, environmental sustainability should be improved. The economic impact of the switch to greener energy will depend on local circumstances, but there have been worldwide declines in the cost of renewable energy over the last decade (https://ourworldindata.org/cheap-renewables-growth).

Example 1B considers the impact of generating electricity from renewable sources on site, The social and societal impacts are similar to those in example 1A. Installing and maintaining a generation facility on site is likely to add to operational complexity, but environmental sustainability will be increased, and there is certainty over the green credentials of the electricity generated. Economically, it will be critical to examine how long it will take for savings in electricity bills to pay back the cost of installation.

9.2 *Example 2: Installing LED Lighting*

Example 2 in Table 2 models a situation where LEDs replace older technologies without changing the level of illumination, length of exposure or other parameters. In the long term, environmental and economic sustainability are enhanced through greater energy efficiency, once the initial cost and embodied carbon of new systems are offset, while operational sustainability is largely unaffected. Social and societal sustainability are unchanged, although some report that light levels can be lowered without reducing visibility (e.g., Padfield et al. 2013) offer the prospect of improving either or both social and societal sustainability by increasing illuminance without causing more damage or decreasing exposure to benefit preservation. This case study is also presented in greater depth elsewhere (Saunders 2022).

Table 2 Analysis of the effects on the five types of sustainability defined in Fig. 2 when a series of scenarios described in the text are applied [This table to be reproduced in colour so that colours match Fig. 2]

Scenario	Category of sustainability	Impact	Notes
1A Switching to a 'green' electricity supplier*	Environmental	⇑	Reduced greenhouse gas emission
	Social	⇔	Access unaffected
	Societal	⇔	Collection lifetime unaffected
	Operational	⇔	Operations unaffected
	Economic	⇓	Potential energy cost implications
1B On-site generation of electricity*	Environmental	⇑	Reduced greenhouse gas emission
	Social	⇔	Access unaffected
	Societal	⇔	Collection lifetime unaffected; effect on building
	Operational	⇓	Increased complexity of on-site generation
	Economic	⇔	Initial investment balanced by long term savings
2 Switch to LED light sources*	Environmental	⇑	Reduced greenhouse gas emission & resource use
	Social	⇔	Access unaffected
	Societal	⇑	Lifetime unaffected
	Operational	⇔	Operations largely unaffected after installation
	Economic	⇑	Reduced energy & replacement costs offset investment
3 Increased use of daylight*	Environmental	⇑	Reduced greenhouse gas emission
	Social	⇔	Access largely unaffected
	Societal	⇔	Lifetime largely unaffected if well managed
	Operational	⇓	Increased complexity: monitoring & maintenance
	Economic	⇑	Reduced energy costs offset any initial investment
4 Switching display lighting 'on demand'	Environmental	⇑	Reduced greenhouse gas emission
	Social	⇔	Access unaffected

(continued)

Table 2 (continued)

Scenario	Category of sustainability	Impact	Notes
	Societal	⇑	Lifetime increased
	Operational	⇓	Potential increased operational complexity
	Economic	⇑	Reduced energy costs offset initial investment

* The symbols ⇑, ⇓ and ⇔ give simplified indications of whether net changes are beneficial, detrimental or neutral, respectively.

9.3 Example 3: Daylighting

Increasing the proportion of daylighting is modelled in Example 3 in Table 2. There will clearly be environmental benefits if this results in a reduction in non-renewable energy and resource use in lighting systems and relamping. Against this might need to be offset the embodied carbon in any daylighting control systems and the energy costs for their operation, although there is likely to be a net increase in environmental sustainability, particularly over the longer term. Broadly, the economic implications will follow the same pattern, with some initial costs for additional daylight control likely offset by longer-term savings. The operational sustainability of increased daylight use will depend strongly on the way in which its variability is managed. If changes to the level of daylight are accepted or there are already methods in place to moderate daylight (i.e. the change is simply to allow greater use of daylight) operational sustainability will be largely unchanged. If new systems are introduced their future continuance and upkeep need to be factored into working practice and maintenance schedules, which has implications for future operational sustainability.

Increased daylighting may also affect social and societal sustainability, although any negative aspects can usually be mitigated. Greater reliance on daylight might result in longer periods of poorer lighting—and hence diminished viewer experience—unless the use of electric lights when daylight fails is well co-ordinated. The potential for greater damage to collections from higher light exposure under variable daylight and from the higher ultraviolet content of daylight can be mitigated by good lighting control and filtration respectively.

9.4 Example 4: Changes in Practice

The final example given in Table 2 models the effect of the introduction of switching of display electric lighting 'on demand', using either a visitor-operated switch or a system to detect the visitor. Decreased use of the lights should reduce energy use and, consequently, greenhouse gas emissions and costs. An added benefit will be lower light exposure for vulnerable materials, improving their long-term preservation—mapped here as societal sustainability.

If the switching regime works well, providing good lighting when visitors are present, then visitor experience (social sustainability) should be unaffected. The more complex the system, the greater its effect on operational sustainability; a simple visitor operated switch is easily maintained, while motion of infrared sensor triggered lighting will require greater longer-term maintenance.

10   Conclusion

As part of the move by cultural heritage organisations to reduce their carbon footprint, changes can be made to museum lighting technology and practice that decrease energy and raw material use, improve energy efficiency and place greater emphasis on reuse and recycling. While the economic implications of such changes are usually considered, less attention is often paid to their operational sustainability within the organisation. In addition, to ensure that anthropocentric elements of long-term sustainability are not overlooked, it is useful to test any proposed scenario using a model—such as that presented here—that also examines its impact on individuals and cultures. Although a definition of the elements that might comprise each 'pillar' is given in Fig. 2, these can be adapted to meet the situation and priorities of individual organisations. The examples given here all derive from steps commonly taken to improve the environmental sustainability of museums, but the model can also be used to examine how a proposed change to practice driven by another consideration—for example preservation of the collection—might affect each measure of sustainability.

References

BEIS. 2016. *Building energy efficiency survey: Community, arts & leisure sector, 2014–15*. London: Department for Business, Energy and Industrial Strategy.

BEIS. 2018. *Business energy statistical summary*. London: Department for Business, Energy and Industrial Strategy.

Bintz, Carol, and Paul Bernard. 2014. The art of efficiency. *High Performing Buildings*: 28–34.

Blades, Nigel, John Mardaljevic, Katy Lithgow, Stephen Cannon-Brookes, Lisa O'Hagan, and Sarah McGrady. 2020. Improved daylight management of historic showrooms: A methodology based on detailed recording and analysis. *Studies in Conservation* 65: P18–P24.

BSI (British Standards Institute). 2010. *BS 8887–220:2010: Design for manufacture, assembly, disassembly and end-of-life processing (MADE)—The process of remanufacture. Specification*. London: British Standards Institute.

BSI (British Standards Institute). 2011. *BS 8887–240:2011: Design for manufacture, assembly, disassembly and end-of-life processing (MADE). Reconditioning*. London: British Standards Institute.

Cannon-Brookes, Stephen, Nigel Blades, Katy Lithgow, and John Mardaljevic. 2017. New devel-
    opments in understanding daylight exposure in heritage interiors. In *ICOM-CC 18th trien-
    nial conference preprints*, ed. J. Bridgland, Art. 1505. Copenhagen: International Council of
    Museums.
Jose L., Casamayor, Su, Daizhong, and Zhongmin Ren. 2018. Comparative life cycle assessment
    of LED lighting products. *Lighting Research & Technology* 50: 801–826.
EC (European Commission). 2011. Directive 2011/65/EU of the European parliament and of the
    council of 8 June 2011. *Official Journal of the European Union* 174: 88–110.
EC (European Commission). 2012. Regulations: Commission delegated regulation (EU) No 874/
    2012 of 12 July 2012. *Official Journal of the European Union* 258: 1–20.
EC (European Commission). 2012. Directives: Directive 2012/27/EU of the European parliament
    and of the council of 25 October 2012. *Official Journal of the European Union* 315: 1–56.
EC (European Commission). 2019. Commission delegated regulation (EU) 2019/2015 of 11 March
    2019. *Official Journal of the European Union* 315: 68–101.
EC (European Commission). 2019. Commission regulation (EU) 2019/2020 of 1 October 2019.
    Official Journal of the European Union 315: 209–240.
Energy Star. 2021. About energy star®–2020. www.energystar.gov/sites/default/files/asset/
    document/2021%20About%20ENERGY%20STAR%20Overview%204.12.21%20v1.pdf.
    Accessed 24 Mar 2022.
EucoLight. 2019. Annual report 2019: Making the circular economy a reality for lighting products.
    www.eucolight.org/biennial-report. Accessed 24 Mar 2022.
Hirsch, Peter, and Christian Schempp. 2020. *Categorisation system for the circular economy—A
    sector-agnostic categorisation system for activities substantially contributing to the circular
    economy*. Brussels: Publications Office.
HM Government. 2011. Carbon plan. https://assets.publishing.service.gov.uk/government/uploads/
    system/uploads/attachment_data/file/47621/1358-the-carbon-plan.pdf. Accessed 24 Mar 2022.
IIC, ICCROM, and ICOM-CC. 2021. Joint commitment for climate action in cultural heritage.
    www.iiconservation.org/sites/default/files/news/attachments/11206-joint_commitment_for_cli
    mate_action_in_cultural_heritage_final_for_release_in_oct_2021.pdf. Accessed 24 Mar 2022.
IRENA (International Renewable Energy Agency). 2021. Renewable capacity statistics. www.irena.
    org/publications/2021/March/Renewable-Capacity-Statistics-2021. Accessed 24 Mar 2022.
MLA and ARUP. 2010. Museums & art galleries survival strategies: A guide for reducing operating
    costs and improving sustainability. https://museumdevelopmentnorthwest.files.wordpress.com/
    2012/06/museum-and-gallery-survival-strategy-guide-printable.pdf. Accessed 24 Mar 2022.
Myer, Michael A., and Bruce R. Kinzey. 2010. Demonstration assessment of light-emitting diode
    (LED) accent lighting at the field museum in Chicago, IL. https://www.eere.energy.gov/buildi
    ngs/publications/pdfs/ssl/gateway_field-museum.pdf. Accessed 24 Mar 2022.
NHM (Natural History Museum). 2020. Sustainability report 2019–2020. www.nhm.ac.uk/con
    tent/dam/nhmwww/about-us/sustainability/sustainability-reports/sustainability-report-2019-
    2020.pdf.pdf. Accessed 24 Mar 2022.
Ohno, Yoshi. 2006. Optical metrology for LEDs and solid state lighting. In *Proceedings of SPIE:
    Fifth symposium optics in industry*, ed. Rosas Eric, Cardoso Rocío, Bermudez Juan C., and
    Barbosa-García Oracio, 604625. Santiago De Queretaro, MX: SPIE.
Padfield, Joseph, Steve Vandyke, and Dawson Carr. 2013. Improving our environment. www.nat
    ionalgallery.org.uk/paintings/research/improving-our-environment. Accessed 24 Mar 2022.
Rahman, S.M., Stéphane Pompidou. Mizanur, Thècle. Alix, and Bertrand Laratte. 2021. A review
    of LED lamp recycling process from the 10 R strategy perspective. *Sustainable Production and
    Consumption* 28: 1178–1191.
Saunders, David. 2022. A methodology for modelling preservation, access and sustainability.
    Studies in Conservation 67: 245–252.
Thickett, David. 2016. Managing natural light in historic properties. In *Lights on... Cultural heritage
    and museums*, ed. Homem Paula M., 245–264. Porto: LabCR, FLUP.

US Department of Energy. 2013. Solid-state lighting technology fact sheet: Lifetime and reliability. www.eere.energy.gov/buildings/publications/pdfs/ssl/life-reliability_fact-sheet.pdf. Accessed 24 Mar 2022.

USEIA (US Energy Information Administration). 2003. Public assembly end use.

USEIA (US Energy Information Administration). 2012. Commercial buildings energy consumption survey: Energy usage summary. www.eia.gov/consumption/commercial/data/2012/. Accessed 24 Mar 2022.

Vlachou-Mogire, Constantina, Giulia Bertolotti, Kathryn Hallett, and Kate Frame. 2020. Developing 'smart' solutions for light management for historic collections. *Studies in Conservation* 65: P333–P341.

Museum Pollutants and Preventive Conservation: State of the Art According to Publications Review

M. J. Alcayde and B. de Tapol

Abstract Although the effects of contaminants on museum collections began to be known since the 18th Century, it is during the last 20 years that indoor pollutants have been considered some of the main contributors affecting the quality of the museum environment. The aim of this paper is to propose an overview of the research according publication review that has been done during the last decades concerning indoor pollutants in museums, internal emissions, selection of materials and mitigation in order to develop preventive conservation strategies.

1 Introduction

Much research has been done and many references have been given with regard to pollutants concerning the natural environment or health, emission limits for the industry, or pollution in cities. But these limits are always much higher than the concentration that could affect materials in museums. Nevertheless, the effects of pollutants on museum objects had already been observed at the end of the 18th Century, like the corrosion caused on lead and shells by volatile organic compounds (Watson 1789).

Initially, Thomson, in his book The Museum Environment (Thomson 1986), set a list of standard pollutants, but indoor pollutants, generated inside the museum, were missing. Thanks to the contributions that Hatchfield and Tétreault made at the beginning of the last decade (Hatchfield 2002; Tétreault 2003), it is possible to get a great compilation of the pollutants found at the museum.

Additionally, some studies have been developed concerning the quantification of exposure-effects, such as the research in pollutants and their NOAEL (non-observed

Research carried out during M.J.Alcayde's internship at Museu Nacional d'Art de Catalunya.

M. J. Alcayde · B. de Tapol (✉)
Museu Nacional d'Art de Catalunya (MNAC), Palau Nacional, Parc de Montjuïc, 08038
Sants-Montjuïc, Barcelona, Spain
e-mail: benoit.detapol@museunacional.cat

Á. F. Perles-Ivars et al. (eds.), *Collection Care*, Springer Proceedings in Archaeology and Heritage, https://doi.org/10.1007/978-3-031-85655-6_8

adverse effect level) values (Tétreault 2003). However, for some authors the recommendation is to reach the lowest possible amounts of contaminants because of the unknown damage caused by their interaction with other parameters and factors (Schieweck and Salthammer 2009). In this regard, some mixtures at low-level concentrations appear to be capable of causing more damage than if the materials were exposed to a single isolated one.

May Cassar established guidelines for risk assessment in the museum environment (Blades et al., 2012). Standards and a number of political implementations are also carried out which are focused on preventive conservation to reduce risks and slow down the damage to collections from pollutants or vandalism thereby continuing on the road to sustainability (European Preventive Conservation Strategy. PC Strat—Raphael Programme. European Commission. Vantaa. 21st–22nd Sept 2000).

Pollutants, which cause damage, are included in the family of contaminants, which themselves are included in the large family of substances (BSI Standards 2012). Volatile pollutants (most volatile have less than 6 carbons and a boiling point under 100°C) are not the only chemical agents of deterioration. It is necessary to beware of the materials in direct contact with the artefact (i.e. polymeric materials for storage—covers, padding-, exhibition–boards, coatings, adhesives, sealants-, or protection–wax, varnish-) that may contain ingredients such as plasticizers or non-volatile acids, which could damage the object by migration or reaction (Tétreault 1996).

2   State of the Art for Indoor Pollutants in Research in Literature

Due to the huge number of documents published, the search has been restricted to documents in the English language dated between 2003 and 2019. The first review is based on the work that has been done on this matter after the contributions of Hatchfield and Téreault made in (2002) and (2003), respectively.

We classify studies in 5 groups, such as (1) identification of pollutants and their sources, (2) effects of pollutants on the artworks, (3) assessment of materials for museum purposes—display and conservation/restoration-, (4) risk management, (5) museum environment—ventilation, particulate matter.

In a second step, another limitation is chosen: only papers in peer-reviewed journals. Six journals are considered which are the most recognised in the field of the conservation of heritage: Journal of Cultural Heritage, e-Preservation Science, Studies in Conservation, Journal of the Institute of Conservation, Conservation Science in Cultural Heritage and Journal of the American Institute for Conservation.

The two figures show differences; In the case of the broad systematic search (Fig. 1), it is seen that, with exception of "Museum environment", all the other 4 groups are quite in equilibrium, meaning similar percentage. On the contrary, in Fig. 2, with selected journals, the articles studying and analysing pollutants are the most abundant (more than 80% of the publications).

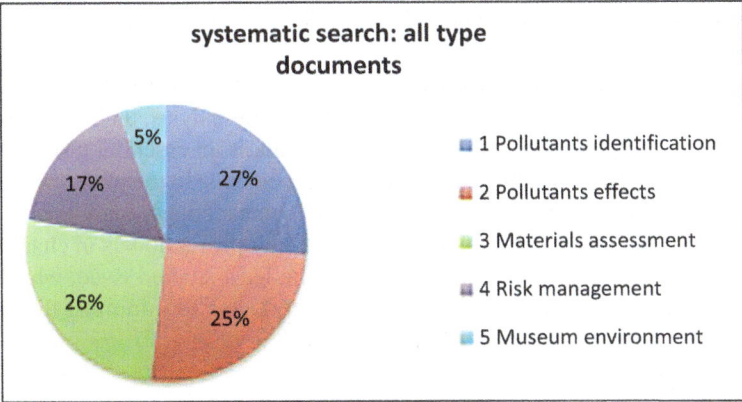

Fig. 1 Result of the systematic search in percentage considering keywords for each group contain in all type of documents in English language: Articles, recommendations, reviews, books and reports

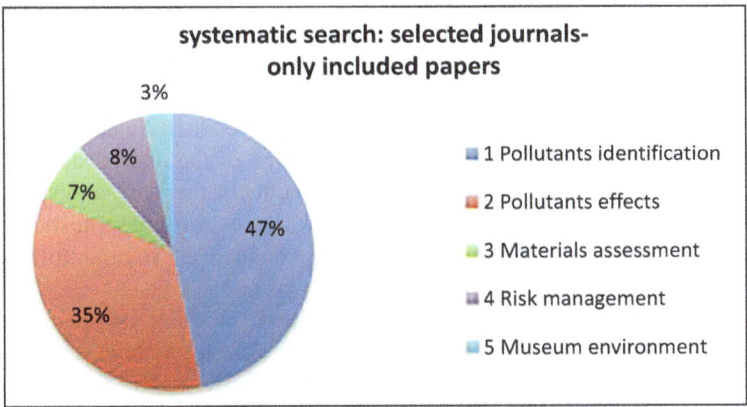

Fig. 2 Results of the systematic search for after revision of key word from selected journals. (*It could be necessary to explain that some articles can be in several groups at the same time due to the interrelation between all those topics.)

Moreover, another remarkable difference is the percentage regarding materials assessment, selection of materials and test procedures for exhibition purpose, only 7% (Fig. 2) in comparison with the 26% in the case of museum documents and newsletters (Fig. 1).

The conclusion is that, during the last 20 years, a great effort has been made in the study of pollutants, their emissions, sources and effects on works of art. There have been developed sensitive analytical techniques, which can quantify the museum contaminants, and monitoring devices or dosimeters to detect the pollutants before they become a real danger to the objects. However, the publications in journals about material assessment, risk management, protocols and standards, and the museum environment are clearly in the minority.

Materials used in contemporary art or materials used in construction or exhibition purposes are changing every day. This means that their analysis, their possible emission, and their compatibility with the objects around, is an essential task that needs to be done continuously for the safeguard of the art objects. In this way, the materials assessment should be improved as well as the risk management. Standards and easy to follow protocols should be known and followed by professionals in charge.

There have been studies focusing on one of the least studied issue, which is the effect of particulate matter, its effects and the strategies to minimize them in the museum environment (Chachula 2021).

3   Airborne Pollutants: Identification, Sources and Effects

It is possible to handle the study of volatile pollutants in the museum environment from two different viewpoints; firstly, the identification and quantification of contaminants and their sources, and secondly, the effects of pollutants on the constituent materials of the artefacts.

Regarding the first point, the methods usually employed for the analysis of contaminants require a certain level of chemical knowledge of the analytical techniques. From the second point of view, for the study of the effects, there are some techniques such as electrochemical, colorimetric, SEM, FTIR, and even more sophisticated ones like synchrotron beam radiation techniques. However, there are some simple methods that are easier to apply, less expensive (Thickett and Lee 2004).

The most sensitive and precise method for identification of pollutants includes active sampling of the air (removing air from the area by manual or mechanical pump) and requires analytical techniques such as GS-MS or HPLC. Detractors of active sampling would say that the environment remains disturbed when sampling (Heesters et al. 2002a).

Some other methods involve collecting pollutants using passive sampling. After the capture of the contaminants with chemical absorbents, analytical chromatographic techniques are used to identify and quantify them. Nowadays, diffusion tubes are employed (Palmes, Dräger, Radiello) even though they are less precise and sensitive and they are easier to apply and more economical (Palmes et al. 1976; Gibson et al. 1997; Gibson and Brokerhof 2001). Some absorbents like Tenax, TEA, KOH and active charcoal could be used to trap pollutants and afterwards be desorbed and identified by GC–MS, HPLC or IC (Heesters et al. 2002a; Stranger et al. 2008; Hodgkins et al. 2011). In some cases, pollutant levels could be indicated by visible colour change that is specific for some individual contaminants, such as formaldehyde. Also, Gibson et al. (2008) developed methods for identifying formaldehyde easily and rapidly with UV–Vis spectroscopy.

All the methods above are qualitative, quantitative and sensitive, but in general, they are quite expensive and time-consuming. There were new ongoing developments in detection devices and dosimeters for monitoring presence of pollutants (Ryhl-Svendsen 2008), some of which are being tested under several European Projects (Odlyha et al. 2000; Bergsten et al. 2010). Special attention should be paid to EU projects (PROPAINT, MEMORI, MUSECORR, APACHE...), which evaluate the response of several dosimeters and monitoring devices (EWO, RM_PQC; GSD, L-PQC, Air Corr logger), their suitability to check pollutants and the degree of the corrosive atmosphere for sensitive materials, in different locations, inside and outside mc-frames (Grøntoft et al. 2010; Dubus and Prosek 2012; Kouril et al. 2013).

The so-called "spot tests" have been developed to determine the suitability of a certain material for its possible use in restoration, conservation or exhibition (Thickett and Lee 2004). Moreover, these tests are simple and cheap and could be carried out easily in the museum or in any laboratory and do not require very specialized staff.

Initially, at the British Museum in 1973, Oddy designed a new method by testing the effect that the emissions of some materials could cause on metal coupons (Oddy 1973; Oddy 1975). The Oddy test is an accelerated corrosion test to evaluate the suitability of products for their use in displays, boxes and cabinets. Since then, many versions have been developed to simplify the test and to evaluate and improve the reproducibility by comparison between different laboratories (Green and Thickett 1995; Bamberger et al. 1999; Robinet and Thickett 2003). This test has been used in many museums and research institutions (Espinoza and Araya 2000; Yoon and Brimblecombe 2000; Kim et al. 2003; Luxford and Thickett 2007; Issaeva and Odegaard 2009). Nowadays, the "3 in one" method developed by Robinet and Thickett in (2003) was currently being used at The British Museum (Robinet and Thickett 2003; Thickett and Lee 2004; Alcayde and de Tapol 2012). For the last version of the Oddy Test in 2021, feedback is not already available (Curran and Strlič 2015).

Oddy Test gives practical results, but the methodology will lead a pour reproducibility (conditions of T°C and RH), without quantification or identification of the chemical compounds (Eremin and Wilthew 1996; Schieweck and Salthammer 2009) conditions of T and RH. So, some experiments have been done to test non-metallic coupons such as acetate cellulose paper (Pretzel 2008) and silver nanoparticle sensor has been developed to evaluate sensibility to Hydrogen Sulphide (Moussa 2007).

Studies for the development of new and rapid methods for the detection of harmful substances and degradation agents for the artwork materials are constantly being published (Tétreault 1996; Maines and Johnston-Feller 1979; Daniels and Ward 1982; Fenn 1995; Nicholson and O'Loughlin 1999; Halsberghe et al. 2005; Matthiesen 2007; Matija et al. 2010) but they are still not very commonly applied by the personnel in charge of the conservation of the collections.

Many studies have been done on the effects caused by VOCs, especially formaldehyde, formic and acetic acid on lead and copper, controlling time, concentration and RH (Tétreault et al. 1998; Raychaudhuri and Brimblecombe 2000; Tétreault et al. 2003; Kraševec et al. 2021) drawing conclusions such as maintaining the level of carbonyl under 0,1 ppm and RH < 40%, avoiding emissive materials and using some other palliative methods, like absorbents or scavengers. Studies on other materials, like calcareous specimens using an artificial environment, have been developed as well (Hodgkins et al. 2008).

Some conclusions by Schieweck are, most VOCs such as alkanes can be regarded as non-reactive, aromatic hydrocarbons cause adverse effects on metal oxide pigments in the presence of NO_x, like on TiO_2; the non-reacted photo-initiators used in the polymerisation of some coatings can lead to corrosion or degradation on near surfaces. And aldehydes may be oxidised to their acids, which can damage sensitive materials (Schieweck and Salthammer 2009). However it was found that formaldehyde has been overestimated as high RH and oxidants are necessary to cause real damage to lead artefacts (Raychaudhuri and Brimblecombe 2000). Furthermore, studies of the materials employed in the construction of showcases have been performed, such as for lacquers and neutral curing sealants (Schieweck and Sakthammer 2011).

Some of the most recent developments are electrochemical methods used to evaluate the effects of VOCs on metals (Costa and Dubus 2007), DSC applications for the parchment degradation (Budrugeac et al. 2010), analysis for the stability of polymeric colours (Fenech et al. 2010b), and analysis of the effects of ozone on photographic material (Burge et al. 2010). This also includes the evaluation of VOCs in archives and libraries and the effects of contaminants on paper (Fenech et al. 2010a; Menart et al. 2011).

Aside from airborne pollutants, some studies on the dust and particles on paintings or in historic houses and museums, and some proposals for its reduction, can also be found (Yoon and Brimblecombe 2000; Lithgow et al. 2005; Alebic-Juretic and Sekulic-Cikovic 2009; Andělová et al. 2010). It was also studied that products obtained by poly-addition, oximes, piperidinol compounds (from UV stabilisers) and others residual manufacturing contaminants of glasses (thorough cleaning) were interacting with typical levels of museum VOCs, developing sodium hydroxide film on them. Periodical removal of these greasy residues is the only solution to slow the formation of crystalline sodium salts (Poulin et al. 2020).

4   About Materials: Emissions and Compatibility

There are studies of the VOCs emissions from coatings, wood and wood products, and other construction materials especially due to their proximity to the artefact in exhibitions or storage and the possibility of creating a damaging microenvironment around the object (Schieweck et al. 2017; Eremin and Wilthew 1996; Schieweck and Salthammer 2009; Paterakis 2016). Studies done on the barrier effect for the moisture and emissions of some coatings applied on wood (Brewer 1991).

From those studies, it can be said that the most effective sealants were considered to be the metallic foils (i.e. Malverseal® 360), although their efficacy could vary depending on the sealing method (Eremin and Wilthew 1996). Another issue would be the aesthetic result of the material that could lead to refusal for exhibition purposes. Deterioration rates may be greatly reduced by implementing careful control of the environment and by understanding the behaviour of the materials.

Tétreault classified them by the type of film forming mechanism (coalescence, oxidative polymerisation, solvent evaporation and catalysed polymerisation) and provided some recommendations about palliative methods and air exchange rates inside the cabinet (Grzywacz and Tennent 1994; Tétreault and Stamatopoulou 1997; Tétreault 1999; Tétreault 2011). At the CCI, the concept of DOSE (concentration of pollutant x duration of the exposure) was adopted. For a certain DOSE of a pollutant, it is possible to calculate the maximum exposure time before signs of damage could appear.

Hence, by 1993, Tétreault published "Guidelines for selecting materials for exhibit, storage and transportation" (Tétreault and Williams 1993) at the CCI. In the end, the fundamental concept that would really be highlighted is compatibility. Materials should not be considered in terms of good or bad, but compatible or not with the artefact. Some other authors share this opinion, paying attention to interaction between different objects due to migration of substances (Valot 1993; Heesters et al. 2002b; Spathis et al. 2003; Ragauskien et al. 2006; Petersen et al. 2008; Matija et al. 2010; Badea et al. 2011).

In general, on that context, new materials such as adhesives and coatings with low VOC content are constantly being developed and controlled (Kremer et al. 2000; Spino 2002; Fine 2007). But even in this new policy (i.e. EN ISO 17895:2005), it would be recommended to test them before use and let them dry for several weeks (3 or 4 weeks) (Tétreault 2011).

Finally, and as already mentioned, compatibility between materials should be respected. Once potential hazards have been determined, it must be ensured that the materials are harmless to the artefacts by controlling the chemical nature of the artwork and environmental factors such as air volume, temperature or humidity.

5 What is Going on in Preventive Conservation: Risk Assessment, Control Strategies

A review has been already published about preventive conservation to evaluate and minimize the risk of wood collections, optimizing display and storage conditions (Luxford et al. 2013). The first goal is to keep outdoor pollutants at a low level inside a showcase or container. For it, it is necessary to have low air exchange and avoid contaminants from being introduced into the cabinets. But showcase, box, and padding or coating material themselves can generate new ones, and in this case high air exchange to dilute them is needed.

Nevertheless, in both cases, they can be reduced by surface absorption that could be increased by adding high and specific absorptive materials. The area of the absorbent product is a crucial factor. Recirculation and filtration of internal air are the most effective mechanisms to minimize contaminants, with the exception of VOCs, as ventilation filters can increase their release. It would be necessary to include some suitable absorbents on filters as well. Combining low air exchange, wall absorbents and recirculation/filtration may be the best solution (Ryhl-Svendsen and Clausen 2009).

It is essential to assess harmless substances, their sources, and their impact on artefacts, as highlighted by Blades and Cassar in their Guidelines on Pollution Control in Heritage Buildings (Blades et al., 2012).

The synergistic effect between pollutants together combined with environmental factors which causes greater damage than the sum of their isolated effects, is also explained. Moreover, other studies provide a contribution to the reduction of pollutants with strategies like the use of scavengers, blocking contaminants or improving air quality by controlling the ventilation and air exchange inside cabinets, showcases and room (Ryhl-Svendsen and Clausen 2009).

However, although the contaminants trapping can be done more or less easily, their individual neutralizations are not always performed. For each of the pollutants, it is necessary to add to the absorbent (or adsorbent material) the specific chemical product that will oxidize, reduce or transform it in order to deactivate it. Activated carbons and zeolite like acid absorbents for the cellulose acetate-based plastics (Ligterink 2002) or for the protection of metals (Cruz et al. 2004), have been developed. Some results alert us to the danger of activated charcoal and zeolites because these sorbents could act as second emission sources. Several specific sorbents have also been tested to check the effect of the acetic acid on calcareous materials under several RH values. The conclusions were that moisture should be as low as possible, provided it does not affect organic materials, and as a second conclusion, that a filter impregnated with KOH would be the best solution for absorbing volatile acids (Brokerhof 2002) Some commercial products, based on potassium permanganate or potassium iodide, are specifically used.

Installing and periodically exchanging activated carbon cloths (or other adsorbent media) inside a museum enclosure was an efficient way to reduce volatile organic acid concentrations in the enclosure.

Recent proposals (results of APACHE project) for capture of gaseous pollutants have emerged, known as "green" biopolymers, like Castor oil (CO)-based gel, or materials based on chitosan (by de-acetylation of chitin extracted from crustaceans), or silica-based composite (substitutes of activated carbon or charcoal).

All these absorbents are inexpensive, recyclable, safe and easy to upscale. They should incorporate specific chemicals to prevent desorption and help to trap variety of pollutants, like zinc oxide (combine with acetic acid) or boron nitride (combine with aldehyde), or PEI polyethyleneimine polymeric species (with e.g. different molecular weights, to capture aldehyde and acid via formation of covalent bonds with the pollutants).

New research has been developed on how pollutants of the twenty-first century could affect our Heritage (Brimblecombe and Grossi 2010). In addition, new studies are using advanced software and technologies, like Computational Fluid Dynamics (CFD) based on the finite elements' method for risk assessment in heritage buildings. This type of work is applied when parameters like variations and interactions between indoor and outdoor microclimatic conditions, and thermo-physical behavior of the building connected to lighting, visitors and HAVC systems; need to be considered for a proper conservation strategy (Balocco 2007).

6 The Use of Cabinets and Showcases: An Improved Control?

One of the strategies to protect artefacts is enclosing objects in cabinets or showcases which could permit an efficient control on RH and be a physical barrier to external pollutants. However, there is an increased potential damage if constituent materials of artefacts emit vapour contaminants inside this confined air volume.

Hatchfield (2002) suggested some questions before using enclosures to protect the artefacts:

- Nature of the objects: emissions, how sensitive it is
- Nature of the pollutants in the environment
- Materials for the enclosures and possible emissions
- Long- or short-term enclosure
- Mitigation tools available

In Propaint European projects http://propaint.nilu.no/Portals/23/PROPAINT-Final%20Report.pdf, studies have been developed in this area of research concerning microclimate frames (Ryhl-Svendsen 2008) for preventive conservation.

Suppliers propose self-compressing magnetic joints for large display cases with glass door on metal. However, many publications study the danger of tightness in a showcase when materials release pollutants and how the effect of the light and the buffering of temperature or humidity can affect to the artworks (Lopez-Aparicio et al. 2006; Bell and McPhail 2007; Shiner 2007; Thickett et al. 2007; Watts et al. 2007; Grøntoft 2010; Grøntoft et al. 2010, Chiantore and Poli 2021).

7 Discussion and Future Perspectives

In the last decade, research dealing with identification, assessment and managing of indoor environmental risks have permit to improve preventive conservation strategies. Through European projects and CEN initiatives, university departments and research laboratories work together with museums, galleries and archives to achieve recommendations to attain an appropriate environment for the conservation of collections. Supporting this idea, new European standards have been developed and implemented to evaluate Relative Humidity or Moisture content in Hygroscopic Materials (i.e. EN 15,757:2010; EN 15,758:2010; EN 16,242:2012) others are in process (Camuffo and

Bertolin 2012; Camuffo et al. 2014); also EU Projects, like ENSEMBLES, CULT-STRAT and Noah's Ark Project, are focus on the study of pollution and climate change effects on Heritage (De la Fuente et al. 2013).

Concerning the Identification of risk, as referred in previous sections, it is necessary to understand the degradation process of the sensitive and hypersensitive materials, to specific contaminants and critical concentration able to start a corrosion process. Although, a conflictive point is the concept of DOSE in identifying situations at risk because pollutants do not appear isolated. They interact with other contaminants and environmental parameters.

For this purpose, it is of great interest to study interactions between outdoor and indoor pollutants, as it has been addressed in recent research as in MEMORY European project (Horizon 2020). A dosimeter has been designed in order to predict the risk when synergies between external pollutants (O3 and NO_x) and internal emanations (acetic and formic acid) and their concentration, are in contact with leather, lead, paper or silver, among others (Memori 2014). The visualization of the results is allowed locating a point in a dyad of three colours: green for the concentrations without risks, orange to indicate a possible risk and red for a real risk. Despite the success, it seems that the commercialization of the Memory's dosimeter ran into.

From our point of view, the way to implement the Risk Assessment should be to develop an easy to follow protocol to select materials for exhibition purposes, but also to select products for remedial conservation processes. In the same way, recent developments of low-cost monitoring instruments of the main pollutants are used to rank and prioritize conservation needs and mitigate the most relevant ones in a cost-effective manner before damaging effects are observed. (Canosa and Norrehead 2019).

One of the "easy tests" for the museum risk assessment is the Oddy test. Victoria and Albert Museum has opened a new way for going further, evaluating the effects on some non-metallic coupons. It could be interesting to continue proposing new coupons and protocols to cover a larger range of materials.

It is fundamental, not only to evaluate the material of the enclosure, their primary or secondary emission, the effect of light on them, but to evaluate if the kind of object will support confined spaces. Works on bad quality of paper and board, photographs, leathers, stuffed animals, most woods, rubbers and plastics, ceramic badly rinsed are some examples of objects with possible dangerous emanations, in the same way that fresh acid-cured materials (such as joins, painting recovering) (Winther et al. 2015). Permanent closed spaces, sealed enclosures and cabinets should be considered only when necessary, adding all the extra precautions like adsorbents, leakage or air recirculation, to avoid the accumulation of internal pollutants. In this way, there is still a long way to go regarding absorbents of contaminants and moisture inside showcases.(Smedemark 2020).

Regarding risk management, two recent risks need to be considered:

- Facing the climate change and how this is going to affect the museum environment. Avoiding accumulation of VOC's by air renovation as it is done currently, represents an energy penalty.
- Museum visitors, who are becoming more and more numerous, as pollutant sources (VOC, dust). Some studies and Museums are considering this aspect already.

A good HVAC with air filtration would be essential to keep a good environment. It is required to avoid possible synergy between contaminants coming from external sources (people, dust, cities pollution), and the ones generated inside the museum. Better filters like porous hybrid materials as MOF's (Metal Organic Framework) represents may be a chance because they are based on green sustainable materials.

It would be useful to regulate the products supplied by the Industry for museum uses in remedial conservation. For instance, their technical data sheets do not often include all the information required. Sometime changes in formulation and/or application could disturb or avoid achieving all of their final properties. The creation of quality labels for materials, with a better characterisation of the products, would be a great help. Even though, some extra tests need to be carried out for verification.

Finally, it is of huge importance to make an effort for a proper diffusion and explanation of the advances achieved in conservation research to the professionals involved in this field. It is also important to avoid confusion about materials and its emissions, to take into account priorities and to provide some new options and strategies in order to help decision makers.

8   Conclusions

This paper presents a compilation of most of the issues in Preventive Conservation, focused in pollutant mitigation, discussed during the last decade. It tries to make an effort to contribute, with a critical analysis, to a better understanding of what is already performed and what could be considered in future research, taking into account the current situation in the investigation institutions and museums, where budgets for research are decreasing.

It's clear that as important as the research itself, it becomes essential the communication of the results, spreading the generated knowledge to actors in conservation. If the information is not read because the language is not inaccessible, effort of communication should be done. Sharing this knowledge will have an impact to perform a proper risk assessment. It is also the role of the national conservation centres, which have laboratories, to publish the results of the tests and the control of quality of the materials used in storage or exhibition, every three years.

References

Alcayde, María J., and Benoit, de Tapol. 2012. Evaluación de las pinturas empleadas en vitrinas de exposición: Emanaciones y efecto barrera. In *Actas del V congreso del GE-IIC, Criterios de calidad en intervenciones*, Madrid.

Alebic-Juretic, Ana, and Sekulic-Cikovic. Duska. 2009. The impact of air pollution on the paintings in storage at the museum of modern and contemporary art, Rijeka, Croatia. *Studies in Conservation* 54: 49–57.

Andělová, Ludmila, Smolík Jiří, Ondráčková Lucie, Ondráček Jakub, López-Aparicio. Susana, Grøntoft. Terje, and Stankiewicz Jerzy. 2010. Characterization of airborne particles in the baroque hall of the national library in Prague. *E-Preservation Science* 7: 141–148.

Badea, Elena, Gatta G. Della, and Budrugeac Petru. 2011. Characterisation and evaluation of the environmental impact on historical parchments by differential scanning calorimetry. *Journal of Thermal Analysis and Calorimetry* 104: 495–506.

Balocco, Carla. 2007. Daily natural heat convection in a historical hall. *Journal of Cultural Heritage* 8: 370–376.

Bamberger, Joseph A., Ellen G. Howe, and Wheeler George. 1999. A variant Oddy test procedure for evaluating materials used in storage and display cases. *Studies in Conservation* 44: 86–90.

Bell, Nancy, and McPhail David. 2007. Managing change: Preserving history. *Materials Today* 10: 50–56.

Bergsten, Carl-Johan., Odlyha Marianne, Jakiela Slawomir, Slater Jonathan, Cavicchioli Andrea, D.L. de Faria, Niklasson Annika Araújo, Svensson Jan-Erik, Bratasz Lukasz, Camuffo Dario, Valle A. Della, Baldini Francesco, R. Falciai, Mencaglia Andrea, F. Senesi, and Theodorakopoulos Charis. 2010. Sensor system for detection of harmful environments for pipe organs (SENSORGAN). *E-Preservation Scince* 7: 116–125.

Blades, Nigel, Tadj Oreszczyn, Bill Bordass, and May Cassar. 2012. *Guidelines on pollution control in heritage buildings*. The council for museums, archives and libraries. Museum Practice. https://discovery.ucl.ac.uk/2443/1/2443.pdf. https://doi.org/10.4324/9781315888484

Brewer, J.A. 1991. Effect of selected coatings on moisture sorption of selected wood test panels with regard to common panel painting supports. *Studies in Conservation* 36: 9–23.

Brimblecombe, Peter and Carlota M. Grossi, 2010. Potential damage to modern building materials from 21st Century Aire P, n°10, 16–24.

Brokerhof, A.W., 2002. Application of sorbents to protect calcareous materials against acetic acid vapors. Contribution to conservation. In *Research in conservation at the Netherland Institute for conservation Heritage*, p 2016.20

BSI, British Standards Institution. 2012. PAS 198:2012: Specifications for managing environmental conditions for cultural collections. London: BSI Standards.

Budrugeac, Petru, Badea Elena, Gatta G. Della, Miu Lucretia, and Comănescu Alina. 2010. A DSC study of deterioration caused by environmental chemical pollutants to parchment, a collagen-based material. *Thermochimica Acta* 500: 51–62.

Burge, Daniel, Gordeladze Nino, Bigourdan Jean-Louis, and Nishimura Douglas. 2010. Effects of ozone on the various digital print technologies: Photographs and documents. *Journal of Physics: Conference Series* 231: 012001.

Camuffo, Dario, and Bertolin Chiara. 2012. Towards standardisation of moisture content measurement in cultural heritage materials. *E-Preservation Science* 9: 23–35.

Camuffo, Dario, Bertolin Chiara, Bonazzi Achille, Campana Francesca, and Merlo Curzio. 2014. Past, present and future effects of climate change on a wooden inlay bookcase cabinet: A new methodology inspired by the novel European Standard EN 15757:2010. *Journal of Cultural Heritage* 15: 26–35.

Canosa, Elyse, and Sarra Norrehed. 2019. Strategies for pollutant monitoring in museum environment. Visby Sweden, https://rgdoi.net/10.13140/RG.2.2.24172.00640 (Date 22.09.2021)

Cassar. May, 2013. Environmental management guidelines for museums and galleries. Routledge. https://doi.org/10.4324/9781315888484

Chachula O. 2021. Rewiew of cost training school: Indoor air quality in museums, galleries and archives in e-conservation Science, 9: 13–17

Chiantore, Oscar, Tommaso Poli. 2021. *Indoor air quality in museum display cases: Volatile emissions, materials contributions, impacts no, in atmosphere*, 12(3): 364. https://doi.org/10.3390/atmos12030364

Costa, Virginia, and Dubus Michel. 2007. Impact of the environmental conditions on the conservation of metal artifacts: An evaluation using electrochemical techniques. In *Museum microclimates: Contributions to the Copenhagen conference, 19–23 November*, eds. Padfield T, Borchersen K, 63–65. Copenhagen, Denmark: The National Museum of Denmark.

Cruz, Antonio Joao, Joao Pires, Ana P. Carvalho, and M. Brotas de Carvalho. 2004. Adsorption of acetic acid by activates carbons, zeolites and others adsorbent materials related with the preventive conservation of lead objects in museum showcases. Journal of Chemical & Engineering Data, 49: 725–731.

Curran, Katherine, and Strlič Matija. 2015. Polymers and volatiles: Using VOC analysis for the conservation of plastic and rubber objects. *Studies in Conservation* 60: 1–14.

Daniels, Vincent, and Ward Susan. 1982. A rapid test for the detection of substances which will tarnish silver. *Studies in Conservation* 27: 58–60.

De la Fuente, Daniel, Jesús Manuel Vega, Fernando Viejo, Iván Díaz, and Manuel Morcillo. 2013. Mapping air pollution effects on atmospheric degradation of cultural heritage. *Journal of Cultural Heritage* 14: 138–145.

Dubus, Michel, and Prosek Tomas. 2012. Standardized assessment of cultural heritage environments by electrical resistance measurements. *E-Preservation Science* 9: 67–71.

Eremin, Katherine, and Paul Wilthew. 1996. The effectiveness of barrier materials in reducing emissions of organic gases form fibreboard: Results of preliminary tests. Vol I. Preventive conservation. ICOM committee for conservation.

Espinoza, Fanny, and Araya Carolina. 2000. Análisis de materiales para ser usados en conservación de textiles. *Conserva* 4: 45–55.

Fenech, Ann, Matija Strlič, Irena Kralj Cigić, Alenka Levart, Lorraine T. Gibson, Gerrit De Bruin, Konstantinos Ntanos, Jana Kolar, and May Cassar. 2010a. Volatile aldehydes in libraries and archives. *Atmospheric Environment* 44: 2067–2073.

Fenech, Ann, Strlič Matija, Degano Ilaria, and Cassar May. 2010b. Stability of chromogenic colour prints in polluted indoor environments. *Polymer Degradation and Stability* 95: 2481–2485.

Fenn, Julia. 1995. The cellulose nitrate time bomb: Using sulphonephthalein indicators to evaluate storage strategies. In *From marble to chocolate: The conservation of modern sculpture*, ed. J. Heuman, 87–92. London: Archetype.

Fine, Harry M. 2007. System and method of applying architectural coatings and apparatus therefor. US Patent 20070196680.

Gibson, L.T., B.G. Cooksey, D. Littlejohn, and N.H. Tennent. 1997. A diffusion tube sampler for the determination of acetic acid and formic acid vapours in museum cabinets. *Analytica Chimica Acta* 341: 11–19.

Gibson, L.T., and A.W. Brokerhof. 2001. A passive tube-type sampler for the determination of formaldehyde vapours in museum enclosures. *Studies in Conservation* 46: 289–303.

Gibson, Lorraine, Kerr William, A. Nordon, J. Reglinski, C. Robertson, L. Turnbull, C.M. Watt, A. Cheung, and Johnstone Walter. 2008. On-site determination of formaldehyde: A low cost measurement device for museum environments. *Analytica Chimica Acta* 623: 109–116.

Green, Lorna R., and Thickett David. 1995. Testing materials for use in the storage and display of antiquities—a revised methodology. *Studies in Conservation* 40: 145–152.

Grøntoft, Terje, Odlyha Marianne, Mottner Peter, Dahlin Elin, Lopez-Aparicio. Susana, Jakiela Slawomir, Scharff Mikkel, Andrade Guillermo, Obarzanowski Michal, Ryhl-Svendsen. Morten, Thickett David, Hackney Stephen, and Wadum Jørgen. 2010. Pollution monitoring by dosimetry and passive diffusion sampling for evaluation of environmental conditions for paintings in microclimate frames. *Journal of Cultural Heritage* 11: 411–419.

Grøntoft, Terje. 2010. Derivation of a model for the calculation of impact loads of air pollutants to paintings in microclimate frames. *E-Preservation Science* 7: 132–140.

Grzywacz, Cecily M., and Norman H. Tennent. 1994. Pollution monitoring in storage and display cabinets: Carbonyl pollutant levels in relation to artifact deterioration. In *Preventive conservation: Practice, theory and research*, ed. A. Roy and P. Smith, 165–170. London: International Institute for Conservation of Historic and Artistic Works.

Halsberghe, Lieve, David Erhardt, Lorraine T. Gibson, and Konrad Zehnder. 2005. Simple methods for the identification of acetate salts on museum objects. In *Preprints of the 14th triennial meeting ICOM committee for conservation, the Hague, 12–16 September 2005*, 639–645.

Hatchfield, Pamela. 2002. *Pollutants in the museum environment: Practical strategies for problem solving, exhibition and storage*. London: Archetype Publications.

Heesters, Raymond, Henk Van Keulen, and Agnes W. Brokerhof. 2002a. Badges for passive monitoring of formaldehyde concentrations in air. In *Contribution to conservations research in conservation at the Netherlands institute for cultural heritage (ICN)*, 25–33. London: James & James.

Heesters, Raymond, Van Keulen Henk, and G.T. Roelofs Wilma. 2002b. Natural resins, artificially aged in steps. In *Research in conservation at the Netherlands institute for cultural heritage (ICN)*, ed. J.A. Mosk and N.H. Tennent, 55–63. London: James & James.

Hodgkins, Robyn E., Cecily M. Grzywacz, and Robin L. Garrell. 2011. An improved ion chromatography method for analysis of acetic and formic acid vapours. *E-Preservation Science* 8: 74–80.

Hodgkins, Robyn, Robin Garrell, and David A. Scott. 2008. Determination of acetic and formic acid concentrations in model systems and identification of efflorescence on calcareous specimens. In *15th triennial meeting ICOM committee for conservation*, ed. Bridgland J, 885. New Delhi: Allied Publishers Pvt. Ltd.

Horizon. 2020. Poem—Participatory memory practices. (s.f.). Research. POEM Horizon. Recuperado el 14 de mayo de 2025, de https://www.poem-horizon.eu/research/

Issaeva, S, and Odegaard N. 2009. *Oddy test: Testing construction materials for the pottery vault, the interpretive gallery and the new conservation lab*. Arizona State Museum Conservation Lab.

Kim, Myoung-Nam, Yu Hei-Sun, and Lee Sung-Eun. 2003. *A small chamber test and Oddy test on medium density fiberboard GRADE (E0, E1). Indoor air quality in museums and historic properties*. Norwich: University of East Anglia.

Kouril, Milan, Prosek Tomas, Scheffel Bert, and Degres Yves. 2013. Corrosion monitoring in archives by the electrical resistance technique. *Journal of Cultural Heritage* 15: 99–103.

Kraševec, Ida, Menart Eva, Strlič Matija, and Kralj Cigić Irena. 2021. Validation of passive samplers for monitoring of acetic and formic acid in museum environments. *Heritage Science* 9: 1–10.

Kremer, W., E. Lühmann, M. Melchiors, and R. Roschu. 2000. Sistemas hidrodiluibles de poliuretano de 2C y UV de baja emisión de VOC para el barnizado de la madera. *Pinturas y Acabados ART* 832: 36–42.

Ligterink, Frank J. 2002. *Notes on the use of acid absorbents in storage of cellulose acetate based materials, Competitive absorption of water and acetic acid on zeolite 4A*. Researche in Conservation at the netherlands Institute of Cultutral Heritage (ICN) 64–73

Lithgow, Katy, Helen Lloyd, Peter Brimblecombe, Young Hun Yoon, and David Thickett. 2005. Managing dust in historic houses–a visitor/conservator interface. In *14th triennal meeting the Hague preprints Vol II preventive conservation ICOM committee for conservation*, 662–669. The Hague: James & James/Earthscan.

Lopez-Aparicio, Susana, Grøntoft. Terje, and Dahlin Elin. 2006. Air quality assessment in cultural heritage institutions using EWO dosimeters. *E-Preservation Science* 7: 96–101.

Luxford, Naomi, and Thickett David. 2007. Preventing silver tarnish—lifetime determination of cellulose nitrate lacquer. In *Metal 07: Proceedings of the ICOM-CC metal working group interim meeting, Amsterdam, 17–21 September 2007*, eds. Degrigny C, Van Langh R, Joosten I, Ankersmit B, 88–93. Amsterdam: Rijksmuseum.

Luxford, Naomi, Strlič Matija, and Thickett David. 2013. Safe display parameters for veneer and marquetry objects: A review of the available information for wooden collections. *Studies in Conservation* 58: 1–12.

Maines, Christopher, and Johnston-Feller Ruth. 1979. Choosing materials for prolonged proximity to museum objects. In *American institute for conservation 7th annual meeting, Toronto*, 44–49. Washington, DC: American Institute for Conservation of Historic and Artistic Works.

Matija, Strlič, Cigić Irena, Alenka Možir, Thickett David, De Bruin Gerrit, Kolar Jana, and May Cassar. 2010. Test for compatibility with organic heritage materials: A proposed procedure. *E-Preservation Science* 7: 78–86.

Matthiesen, Henning. 2007. A novel method to determine oxidation rates of heritage materials in vitro and in situ. *Studies in Conservation* 52: 271–280.

Memori. 2014. *Measurement, effect assessment and mitigation of pollutant impact on movable cultural assets.* Innovation Reseach for Market transfer. Final report Summary. European Commission. https://cordis.europa.eu/project/id/265132/reporting (Date 22.09.2021)

Menart, Eva, De Bruin Gerrit, and Strlič Matija. 2011. Dose–response functions for historic paper. *Polymer Degradation and Stability* 96: 2029–2039.

Laura Moussa. 2007. *Nanoscience and nanotechnology applied to art conservation: Inproved Oddy test using silver nanoparticle sensor.* Mellon college of science. The Carnegie Mellon University. Thesis

Nicholson, C., and C.E. O'Loughlin. 1999. Screening conservation, storage, and exhibit materials using acid-detection strips. *Collections Caretaker Northern States Conservation Center* 1: 4–5.

Oddy, William A. 1973. An unsuspected danger in display. *Museums Journal* 73: 27–28.

Oddy, William A. 1975. *The corrosion of metals on display. Conservation in archaeology and the applied arts.* London: IIC.

Odlyha, Marianne, Gary M. Foster, Neil S. Cohen, Sitwell Christine, and Bullock Linda. 2000. Microclimate monitoring of indoor environments using piezoelectric quartz crystal humidity sensors. *Journal of Environmental Monitoring* 2: 127–131.

Palmes, E.D., A.F. Gunnison, J. Dimattio, and C. Tomczyk. 1976. Personal sampler for nitrogen dioxide. *American Industrial Hygiene Association Journal* 37: 570–577.

Paterakis, Alice B. 2016. *Volatile organic compounds and the conservation of inorganic material.* London: Archetype Publication. https://doi.org/10.1080/19455224.2017.1416964 (Date 22.09.2021).

Petersen, K., C H. Heyn, and Wolfgang E. Krumbein. 2008. Degradation of synthetic consolidants used in mural painting restoration by microorganisms. In *Les anciennes restaurations en peinture murales*, 47–58. Dijon: Journées d'Études de la SFIIC.

Poulin, J., H. Coxon, J.R. Anema, K. Helwig, and M.C. Corbeil. 2020. Investigation of fogging on glass display cases at the Royal Ontario Museum. *Studies in Conservation* 65 (1): 1–13. https://doi.org/10.1080/00393630.2019.1674479(Date22.09.2021).

Pretzel, B. 2008. *Standard Materials for corrosiveness teesting.* London: Victoria &Albert Museum Journal, n°43 https://www.vam.ac.uk/res_cons/conservation/journal/number_43/corrosiveness. 21/07/2008

Ragauskien, Daina, Niaura Gediminas, Matulionis Eimutis, and Makuška Ričardas. 2006. Long-term and accelerated ageing of an acrylic adhesive used as a support for museum textiles. *Studies in Conservation* 51: 57–68.

Raychaudhuri, Michele R., and Brimblecombe Peter. 2000. Formaldehyde oxidation and lead corrosion. *Studies in Conservation* 45: 226–232.

Robinet, Laurianne, and Thickett David. 2003. A new methodology for accelerated corrosion testing. *Studies in Conservation* 48: 263–268.

Ryhl-Svendsen, Morten, and Clausen Geo. 2009. The effect of ventilation, filtration and passive sorption on indoor air quality in museum storage rooms. *Studies in Conservation* 54: 35–48.

Ryhl-Svendsen, Morten. 2008. Corrosivity measurements of indoor museum environments using lead coupons as dosimeters. *Journal of Cultural Heritage* 9: 285–293.

Schieweck, A., Markewitz D., and Sakthammer, T., 2017. Screening emission analysis of construc-
    tion materials and evaluation of airborn pollutants in newly constructured display cases. *Museum
    microclimates. National Museum of Denmark*, 67–72.
Schieweck, A., and Sakthammer, T. 2011. Indoor air qualityin passive-type museum showcase.
    Journal of Cultural heritage, 12: 205–2113.
Schieweck, Alexandra, and Salthammer Tunga. 2009. Emissions from construction and decoration
    materials for museum showcases. In *Studies in conservation*, 54, 2 (4): 218–235. https://www.
    tandfonline.com/doi/abs/10.1179/sic.2009.54.4.218
Shiner, J. 2007. Trends in microclimate control of museum display cases. In *Museum microclimates*,
    eds. Padfield T, Borchersen K, 267–275. Copenhagen, Denmark: National Museum of Denmark.
Smedemark, Signe H., Morten Ryhl-Svendsen, and Jorn Toftum. 2020. Distribution of temperature,
    moisture and organic acids in storage facilities with heritage collections. *Building Environment*.
Spathis, Panayotis, Karagiannidou Evi, and Magoula Anastasia-Eleni. 2003. Influence of titanium
    dioxide pigments on the photodegradation of paraloid acrylic resin. *Studies in Conservation* 48:
    57–64.
Spino, A. 2002. *Les Panneaux écolos*. La maison du 21e siècle.
Stranger, Marianne, Potgieter-Vermaak. Sanja, Sacco Paolo, Quaglio Franco, Pagani Diego, Cocheo
    Claudio, Godoi Ana Flavia. Locateli, and Van Grieken René. 2008. Analysis of indoor
    gaseous formic and acetic acid, using radial diffusive samplers. *Environmental Monitoring and
    Assessment* 149: 411–417.
Tétreault, Jean, and Stamatopoulou Eugénie. 1997. Determination of concentrations of acetic acid
    emitted from wood coatings in enclosures. *Studies in Conservation* 42: 141–156.
Tétreault, Jean, and Williams Scott. 1993. *Guidelines for selecting materials for exhibit, storage
    and transportation. Version 4.3*. Ottawa: CCI.
Tétreault, Jean, Cano Emilio, Van Bommel Maarten, Scott David, Dennis Megan, Barthés-
    Labrousse. Marie-Geneviève, Minel Léa, and Robbiola Luc. 2003. Corrosion of copper and
    lead by formaldehyde, formic and acetic acid vapours. *Studies in Conservation* 48: 237–250.
Tétreault, Jean, Sirois Jane, and Stamatopoulou Eugénie. 1998. Studies of lead corrosion in acetic
    acid environments. *Studies in Conservation* 43: 17–32.
Tétreault, Jean. 1996. La mesure de l'acidité des produits volatils. *Journal of the International
    Institute for Conservation—Canadian Group* 17: 17–25.
Tétreault, Jean. 1999. *Coatings for display and storage in museums*. In Technical Bulletin n°21,
    Ottawa: Canadian Conservation Institute.
Tétreault, Jean. 2003. *Airborne pollutants in museums, galleries, and archives: Risk assessment,
    control strategies, and preservation management*. Ottawa: CCI.
Tétreault, Jean. 2011. Sustainable use of coatings in museums and archives—some critical obser-
    vations. *E-Preservation Science* 8: 39–48. https://www.morana-rtd.com/e-preservationscience/
    2011/Tetreault-05-01-2011.pdf
Thickett, David, Philip Fletcher, Andrew Calver, and Sarah Lambarth. 2007. The effect of air
    tightness on RH buffering and control. In *Museum microclimates*, eds. Padfield T, Borchersen
    K, 245–251. Copenhagen, Denmark: National Museum of Denmark.
Thickett, David, and Lorna R. Lee. 2004. *The selection of materials for the storage or display of
    museum objects*. London: The British Museum.
Thomson, Garry. 1986. *The museum environment*. London: Butterworths.
Valot, Henri. 1993. A propos des materiaux et de la restauration-dérestauration des peintures
    murales. In *Les anciennes restaurations en peinture murale*, 33–45. Dijon: Journées d'études
    de la SFIIC.
Watson, Richard. 1789. *Chemical essays*. London: T. Evans. https://archive.org/details/chemicale
    ssays01watsgoog
Watts, Siobhan, David Crombie, Sonia Jones, and Yates Sally A. 2007. Museum showcases: Spec-
    ification and reality, costs and benefits. In *Museum microclimates*, eds. Padfield T, Borchersen
    K, 253–260. Copenhagen, Denmark: National Museum of Denmark.

Winther, Thea, Judith Bannerman, Hilde Skogstad, Mats KG Johansson, Karin Jacobson, Johan
    Samuelsson. 2015. Adhesives for adhering polystyrene plastic and their long term effect. *Studies
    in conservation*, 107–120.
Yoon, Young H., and Brimblecombe Peter. 2000. Contribution of dust at floor level to particle
    deposit within the sainsbury centre for visual arts. *Studies in Conservation* 45: 127–137.

Sound-Induced Vibrations as a Potentially Damaging Risk to Museum Collections

Khadija Alami, Christian Baars, Rhys Pullin, and Stephen Grigg

Abstract Previous research on the potentially damaging effects of vibration on cultural heritage items has largely focussed on earthquakes, road and rail traffic, and heavy construction activities. Comparatively little attention has been paid to vibration caused by sound, such as music, but with museums relying increasingly on income from events such as weddings and concerts, this cause of vibration is of growing concern in a heritage context. Anecdotal observation and some publications have reported objects 'wandering' on glass shelves, bumping into each other, and uneasiness about fragile objects potentially being excited by vibration, and consequently being damaged. As part of this study, a data acquisition system was developed which captured both sound and vibrations simultaneously. Sound reaching a display case was analysed and the potential for sound-induced vibrations to be transmitted to heritage items inside the display case was evaluated. It was found that low frequency sound below 200 Hz was the cause of most of the vibration observed. Greater sound levels (louder music) did not necessarily cause greater vibration levels. One way of mitigating potentially negative effects of sound during events in museums is not necessarily to limit sound levels, but to dampen the museum display cases and filter damaging frequencies from music being played.

Keywords Acoustics · Noise · Sound · Vibration · Reverberation · Fatigue · Artefact damage · Low frequency

K. Alami (✉)
Acoustics, Technical University of Denmark, Copenhagen, Denmark
e-mail: alami.k@outlook.com

C. Baars
Department of Collections Care, National Museums Liverpool; Previously: National Museum Cardiff, Cardiff, UK

R. Pullin · S. Grigg
School of Engineering, Cardiff University, Cardiff, UK

© The Author(s) 2025   107
Á. F. Perles-Ivars et al. (eds.), *Collection Care*, Springer Proceedings in Archaeology and Heritage, https://doi.org/10.1007/978-3-031-85655-6_9

1 Introduction

Amongst all sources of vibration potentially affecting cultural heritage items, earth-quakes (cf. Yang et al. 2020; Johnson et al. 2013) have the potential for producing the most dramatic consequences. Most vibration experienced by museum collections is low-level and may be caused by visitor footsteps (cf. Pieraccini et al. 2017), traffic or nearby construction (cf. Thickett 2002; Wei et al. 2018). Vibration may cause physical damage to fragile objects or cause previous conservation work to fail. In the experience of one of the authors (CB), one frequent vibration-related problem is items moving on their shelves, often linked to differential friction. High-profile examples include an Egyptian statuette made of steatite that revolved around its axis whilst on display on a glass shelf at Manchester Museum (Zuanni and Price 2018). Such movements may result in damage if an item knocks into other object on the same shelf, or if it falls off the shelf. At National Museum Cardiff, several incidences of objects moving from their original display position have been recorded over many years. This includes fossil ichthyosaur ophthalmic plates sliding off the glass shelf they were mounted on, and a fossils rhinoceros tooth sustaining damage when it fell off its shelf, resulting in its removal from display for conservation work to be undertaken. Approximately 12 ceramic items in display cases around the balcony in the museum's Main Hall were observed to have moved from their original position following a concert in this space. This event focused the attention of the museums' preventive conservation team on sound, particularly loud music during events held at the museum, as an occasional but potentially major source of vibration. Other staff reported that the motion-sensing security sensors attached to objects and display cases were triggered during events at which loud music was played.

The preventive measures to decrease the effects of vibration on display objects taken by the museum to date include

- The installation of 1 mm thickness silicon discs underneath object mounts to increase friction between the display item and the glass shelf it is mounted on,
- Reduction of the allowable noise levels during events, and music being cut off automatically when it exceeds 85 dB.

Acoustic panels were installed in 2018 between pillars in the museum's Main Hall with the intention of improving the acoustics of that space. While acoustic panels are commonly used to attenuate noise indoors, there is no evidence that they also prevent vibration of any museum displays.

Museum design rarely considers architectural acoustics and sound attenuation. National Museum Cardiff has smooth wall surfaces and high ceilings, increasing the total reflective surface of the inside of the building. One of the architectural aspects of National Museum Cardiff that may increase potential noise-caused vibrations is the central dome which forms part of the building's entrance hall. Domes focus reflected sound, and their symmetrical geometry enables reflections to arrive simultaneously (Obeid 2010), meaning that display cases underneath the dome are likely to receive reflected noise.

Vibration is defined as oscillations of solids that are propagated through wave motion. Vibration may give rise to audible sound which, is defined as oscillations through fluids and may, in turn, cause vibration. Sound may also be defined as an energy which is absorbed, transmitted, and/or reflected when travelling through one medium or more. The quality of vibration absorption, transmission and reflection varies with the frequency and the material it travels through. Both phenomena strongly inter-relate and decay with distance, by energy absorption, but also when they encounter obstacles and discontinuities.

In general, for sound and vibration to interact with a body, the body must possess inertia and elasticity. For sound and vibration to significantly damage a structure, not only do they need to meet its natural resonant frequency f_n, which is the maximum oscillatory response at a specific frequency but they also need to overcome the body's own strength (Marquart 2016). While one singular displacement may be negligible, the same displacement over many cycles may be significant. Cyclic displacements could lead to deformations, cracks, fatigue, and failure of artefacts. By analogy, light, also a form of vibration, causes damage through cumulative discoloration (Wei et al. 2014).

In one previous study, singular vibration (shock) levels which proved to be damaging were reported to be between 0.2 g and 0.6 g (Thickett 2002). An object may vibrate horizontally, potentially causing it to 'walk' off a shelf, or vertically, in which case the artefact may abrade against the surface it rests on. In either case, the vibration intensity must exceed the friction between the artefact and the surface to cause a displacement. In addition, low vibration frequencies may produce structural fatigue and failure, whilst high frequency modes produce noise (Norton 1989).

Some previous pilot projects between National Museum of Cardiff and Cardiff University School of Engineering explored the potential causes for object movements inside display cases, which were largely due to visitor footsteps. This follow-up project sought to determine whether loud music played during events in the museum posed a risk to objects on display.

2 Methodology

Both vibration and sound were recorded inside a display case situated on the north side of the balcony in National Museum Cardiff's Main Hall. Sound and vibration recordings were undertaken inside a large free-standing display case 2100 mm high x 1100 mm wide x 1100 mm deep during separate occasions in December 2019. The display case was made of 12 mm thick glass with insulating rubber seals on the enclosing edges of each frame, and comprised of two vertical levels, each with two shelves 950 mm long x 510 mm deep x 10 mm thick, suspended by steel rods (Fig. 1). The display case sat on a steel plinth 310 mm high.

Two ADXL337 3D accelerometers with a minimum full-scale range of ± 3 g were selected for the vibration measurements and placed on two different levels inside the display case to determine whether one level would be more susceptible to

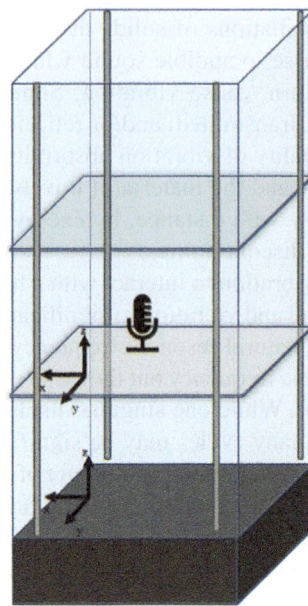

Fig. 1 The museum display case used for the measurements was a large free-standing display case 2100 mm high x 1100 mm wide x 1100 mm deep. The display case was made of 12 mm thick glass with insulating rubber seals on the enclosing edges of each frame. The display case sat on a steel plinth 310 mm high and comprised of two vertical sections, each with two shelves 950 mm long x 510 mm deep x 10 mm thick. The setup inside the museum display case shows a 1/4" free-field microphone, and two three-dimensional accelerometer (a2) on the lower of two vertically arranged glass shelves and an accelerometer (a1) on the plinth

noise compared to the other. Accelerometer a1 was attached to the display case base, and accelerometer a2 was attached to the lower shelf. It was hypothesised that the vibration levels recorded by a1 would be lower than the vibration levels experienced by the display case frame due to the damping effect of the plinth. Nonetheless, the values recorded would be indicative of the vibration direction experienced. The noise reaching the objects in the display case was measured using a G.R.A.S 46BE 1/4" CCP free-field standard microphone optimized for acoustic applications where the location of the main sound source is known. The microphone was placed inside the display case on top of the upper shelf, pointing towards the South side of the main hall where the suspected dominant music source would be located. The National Instruments programming software LabView was used to record and save the data. Mounting the instruments inside the display case had the advantage of avoiding interaction of visitors with the apparatus.

A total recording time of 69 h on different days during a two-week period was used for the data analysis. The recordings were split as shown in Table 1:

- Data set 1 was captured overnight between a Friday evening and a Saturday morning;

Table 1 Summary of recorded sound and vibration data split into three recording sets by sound level: Two loud Christmas parties for the loudest recording set, a busy weekend for a loud recording set and during a quiet weekday for the quietest recording set. A total recording time of 69 h over two weeks was used for the data analysis. The sound threshold was based on the sound level, equivalent to 40 dB

Recording	Duration (hours)	Event
SET 1	15	Quiet day
SET 2	18	Weekend
SET 3	36	Christmas parties

- Data set 2 recorded two Christmas parties;
- Data set 3 was recorded during the weekend of the 7th and 8th December and included a daytime recording with visitors present in the building.

The sound threshold was based on the sound level, equivalent to 40 dB, rather than on the sound frequencies since any potentially relevant frequencies were still unknown at this stage. Any frequencies below 4 Hz were not assessed due to the range limitation of the microphone. A frequency analysis was conducted to investigate the relationship between sound and vibrations. Frequency analysis is a process by which a time-varying signal in the time domain is transformed to its components in the frequency domain. This was implemented digitally using an algorithm known as the Fast Fourier Transform.

3   Results

3.1   Sound

Data set 1 was the quietest of the events recorded; it had the least number of waveforms crossing the threshold of 40 dB (seven waveforms). The lowest peak recorded was 4 Hz during this recording, corresponding to a wavelength of 86 m.

Data set 2 had the most waveforms saved (3913 waveforms). The lowest peak recorded was 4 Hz during the first recording, corresponding to a wavelength of 86 m. The large difference between the number of waveforms crossing the threshold is due to the second recording set being the loudest set as it recorded two Christmas parties.

Data set 3 recorded an intermediate amount of 180 waveforms.

An initial assessment of the sound frequencies indicated that the dominant frequencies observed were mainly below 200 Hz for all recordings. To illustrate this sound pattern, all seven saved waveforms from the data set 1 were plotted in Fig. 2. For clarity purposes, the plot axes were zoomed in to emphasize the dominant peaks. Data sets 2 and 3 produced few significant peaks between 300 and 400 Hz. This was expected considering the display case glass being 12 mm thick, allowing mainly the lower frequencies (longer wavelengths) to be transmitted.

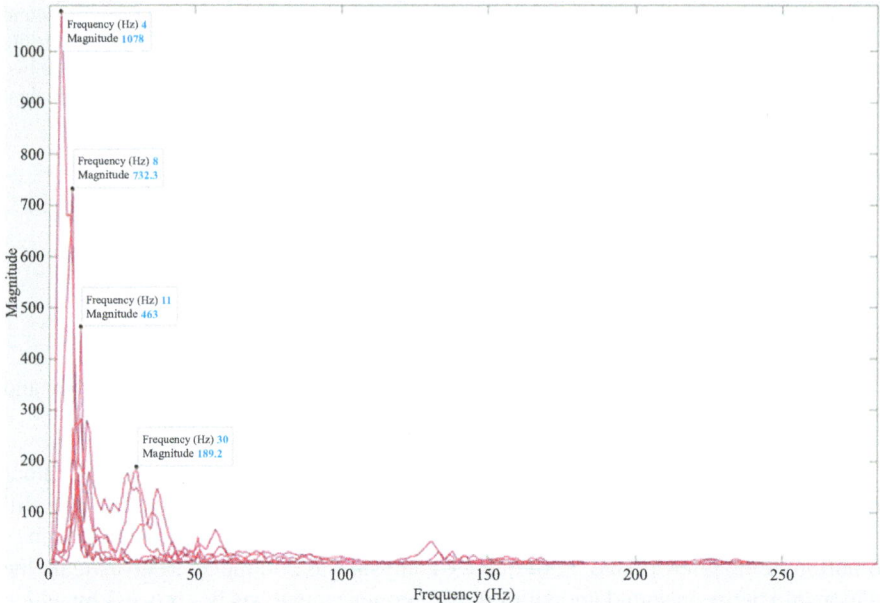

Fig. 2 Example of noise frequencies of the saved waveforms of the 1st recording set. Plot zoomed on the frequency range of 0 to 250 Hz, highlighting low frequency dominant peaks from 4 to 30 Hz. Frequencies lower than 4 Hz were not considered due to the limitations of the microphone

3.2   Vibration

Data set 2, which was the loudest, saved none of the vibration waveforms, meaning the vibration levels were too low. In contrast, the data set 1, the least loud, had the largest number of vibration waveforms (32,396 waveforms). Data set 3 recorded 38 waveforms crossing the thresholds. While there was no direct link between the sound level threshold and the vibration level thresholds, it is still worth noting that both data crossed thresholds inversely. The vibration waveform frequencies displayed a consistent pattern of a peak at 51 Hz, as shown below in Fig. 3, with a dominant frequency of 51 Hz. The maximum vibrations were 0.16 g and 0.2 g in the Z and X axes, respectively.

4   Discussion

The primary areas of concern in this work were the potentially vibration-inducing sound-level and frequency of loud music played during events at the museum. While it is widely perceived that sound-related vibration may be decreased by reducing the noise level, it is important to keep in mind the importance of frequency. The results of

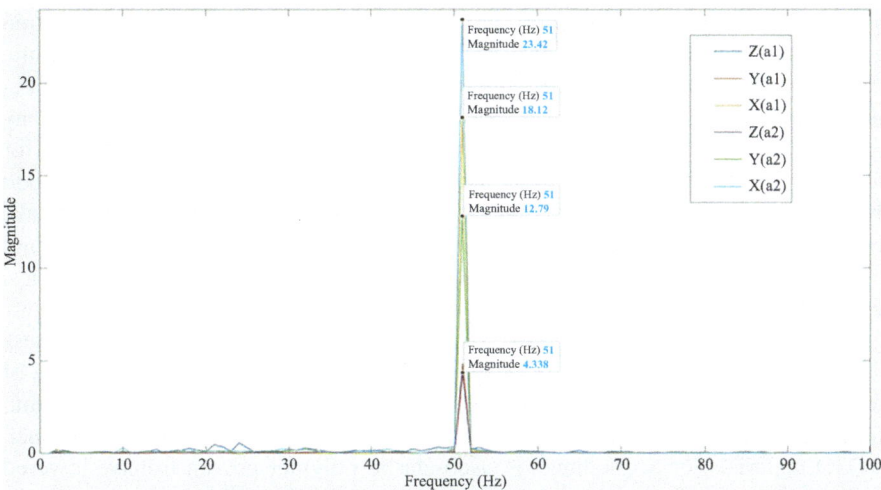

Fig. 3 Example of vibration frequencies from waveforms of the 2nd recording set. Plot zoomed on the frequency range 0 to 100 Hz, highlighting consistent frequency peaks at 51 Hz. Highest magnitude of vibration is recorded in the horizontal direction X by the accelerometer attached to the suspended shelf of the display case (accelerometer a2)

this study demonstrate that the noise reaching objects inside the display case was of low frequency, mainly below 200 Hz. The recording set with the lowest frequencies and highest number of vibrations occurred on a Friday evening when no event was scheduled in the museum and no visitors were present. The very low frequency noise recorded inside the display case may have been due to construction work which, at that time, was undertaken on the roof of the building; electrical appliances, lifts, or mechanical air handling equipment within the building. Alternatively, the low frequencies (Fig. 2) during the first recording on a Friday evening may have been caused by activities at the adjacent Winter Wonderland fun fair, which took place at that time in front of the museum building. Conversely, the recording set with two loud music events (Christmas parties) only recorded background vibration; it is possible that any higher frequencies generated by the music during these events was absorbed by the display case materials. While the thresholds for background sound and vibration levels were chosen independently from each other, it is important to note that both data crossed the thresholds inversely. Hence, there is strong evidence that higher sound levels, such as caused by loud music, do not necessarily cause more vibration.

It is usually preferable to approach noise-related issues at the sound source itself first. This may be achieved by filtering certain frequencies which has the added advantage that, unlike reducing the noise levels, guests would notice no or very little difference in their experience of music—unless bass instruments formed a significant part of the music. If filtering is not possible, noise control may involve building acoustics. When neither are possible, the noise control may also be achieved by enclosure or by sound absorbers. Practical solutions may include the integration of sound absorbers on either the back of the display cases or in the museum building, with the most common being open-cell foams, mineral wool, glass wool, fiberglass, mass loaded vinyl, and polystyrene. While such sound absorbers are effective, many are made of synthetic fibers which produce dust particles that may be as harmful for the environment and museum items as they are for health. Recent research on bamboo fibers, sheep wool, coconut fibers, kenaf fibers, and oil palm mesocarp fibers have proven that more sustainable alternatives may be just as effective, reaching sound absorption coefficients of 0.4 in low frequencies and of 0.8 in higher frequencies, depending on the thickness, density and diameter (Yahya and Chin 2017).

Vibration damping may be achieved by adding a mass or by reducing the stiffness which would lower the natural frequency of the object (Espinosa 2018). Some of the dampening, or cushioning, materials used for sound absorption may also damp vibrations; this includes rubber anti-vibration springs, shock absorbing pads such as D3O materials, or Sorbothane, a visco-elastic polymer proven to have lowered the natural frequency of an artefact and its mount in the British Museum to 1 Hz (Thickett 2002). Display cases and artefacts are impacted by noise and vibration differently, depending on their inertia, geometry, design, and natural frequency; each situation must be assessed independently and it is difficult to give generic advice. In addition, when fitting sound absorption or vibration dampening materials to museum display cases it is also important to consider that many materials produce emissions which may be harmful to cultural history objects and may cause chemical damage (cf. Chiantore and Poli 2021).

All axes in both accelerometers and in every recording, set produced the exact same dominant frequency of 51 Hz. Concludingly, this may have been the accelerometers' or the display case natural frequency. This frequency may be considered potentially damaging to any objects displayed in this display case, especially if their own resonant frequency coincides with this. Wei et al. (2014) focused on frequencies below 50 Hz as they were deemed damaging in his study of allowable vibrations for museum objects. In our study, not only were the vibration levels just as consistent as the frequencies in all three recordings, but their values also coincided with the findings of Thickett (2002). The maximum vibrations of 0.2 g recorded in this present study were of a similar order to those which Thickett (2002) had set as critical and damaging enough to cause abrasion against mounting pins, opening of cracks, and loss of paint flakes. Furthermore, it was observed that the accelerometer on the shelf (a2) recorded greater vibration in the horizontal plane, whereas the accelerometer on the base (a1) recorded greater vertical vibration. This confirms the results presented by Thickett (2002) and Lasyk et al. (2008): different loads may cause a certain level of vibration in a specific direction, highlighting that the surface or support on which an object sits also has an impact on the vibration direction.

One further important aspect is that it may not be adequate to set a critical level for vibration. Based on the definition of vibration itself, the focus should be on the number of cycles, or duration of the exposure of heritage items to vibration, rather than on the intensity of one singular load. Fatigue failure may occur at a much lower stress load than the yield stress of the object (cf. Luxford et al. 2013). This means that smaller cyclic loads, for example, of 0.05 g or even lower, over an extended period might cause similar or more damage than one singular shock load of 0.5 g. Fatigue failure, similarly to light damage, is a cumulative effect. Although setting critical vibration levels would offer easy-to-follow guidance to museums on shock or impact loads, such levels would not necessarily decrease the impact of fatigue failure or object movements if the items experienced many cycles over an extended period of time. Wei et al. (2014, 2017) stressed the need to determine the combination of vibration levels and duration of vibration, or number of cycles at which an object could fail. Hence, this project focused on the frequency range causing the largest number of vibrations, rather than on the frequency range or sound level causing high vibration levels.

5   Conclusions

This project has revealed that the main frequencies experienced on the inside of a museum display cases were below 200 Hz. Loud music was not the only noise source of the vibrations recorded during this study, as there were other (non-identified) sources of low frequency noise at times when no music was played.

On a practical level, reducing sound levels—such as turning down the volume of music—alone is not very effective at decreasing vibration. Instead, vibration-related risk to display objects may be decreased by filtering certain frequencies from music, or by dampening display mounts or cases. Filtering may be achieved by, for example, applying low frequency filters to music at source, but this would not remove vibration caused by mechanical ventilation systems, road traffic or other sources of low frequencies. Dampening involves physical modifications to a display case, shelf or object mount and, hence, may decrease vibration from a range of different sources. Lateral object movements, such as wandering along a glass shelf, may be decreased by increasing friction between the object and the shelf, for example, by mounting the object on a thin silicon sheet.

The natural frequencies of many materials and objects are in the low range, which, as demonstrated in this study, may be less affected by music than by other causes. This means that damage due to the impact of low frequency events at or near the natural frequency of museum objects during occasional events may be less of a concern than cumulative exposure. This does not mean that music does not have the potential to cause vibration-related damage.

There are unresolved questions in relation to the natural frequencies of objects, and there is potential to conduct additional controlled experiments by submitting an object to different frequencies. It would be beneficial if the chosen item had a known natural frequency, as it would be easier to isolate the other frequencies with the greatest impact. A more in-depth analysis may also be achieved by recording sound and vibration at different locations in the museum, because the impact may differ from one display case to another depending on their position relative to the noise source. Measuring microscopical damage of museum objects after or during exposure to sound for an extended period of time may indicate the amount of vibration required to cause damage, and potentially help define acceptable thresholds. From a heritage conservation point of view it would be useful to predict a potential level of damage to an object if all other parameters, including the amount of cyclic loading, were known.

Acknowledgements We are grateful for the assistance of the National Museum Cardiff Preventive Conservation, Curatorial, Technical and Exhibition teams for their permission to use the balcony display case for this study, and their assistance with setting up the experiment.

References

Chiantore, Oscar, and Tommaso Poli. 2021. Indoor air quality in museum display cases: Volatile emissions, materials contributions, impacts. *Atmosphere* 12: 364.

Espinosa, María G. 2018. *Risk estimation of rocking components subjected to ground motions.* PhD Thesis. Oxford, UK: Linacre College, University of Oxford.

Johnson, Arne P., W. Robert Hannen, and Frank Zuccari. 2013. Vibration control during museum construction projects. *Journal of the American Institute for Conservation* 52: 30–47.

Lasyk, Łukasz, Michał Łukomski, Łukasz Bratasz, and Roman Kozłowski. 2008. Vibration as a hazard during the transportation of canvas paintings. *Studies in Conservation* 53: 64–68.

Luxford, Naomi, Matija Strlič, and David Thickett. 2013. Safe display parameters for veneer and marquetry objects: A review of the available information for wooden collections. *Studies in Conservation* 58: 1–12.

Marquart, Sarah. 2016. The science of sound. https://futurism.com/science-sound-breaking-glass-panes-car-speakers. Accessed 1 Mar 2020.

Norton, Michael P. 1989. *Fundamentals of noise and vibration analysis for engineers.* Cambridge, UK: Cambridge University Press.

Obeid, Hani S. 2010. *Control of reverberation times in dome-shaped halls.* Jordan: Applied Sciences University.

Pieraccini, Massimiliano, Michele Betti, Davide Forcellini, Devis Dei, Federico Papi, Gianni Bartoli, Luca Facchini, Riccardo Corazzi, and Vladimir C. Kovacevic. 2017. Radar detection of pedestrian-induced vibrations on Michelangelo's David. *PLoS One* 12: e0174480.

Thickett, David. 2002. Vibration damage levels for museum objects. In *ICOM committee for conservation 13th triennial meeting*, 90–95. Rio de Janeiro: James & James (Science Publishers) Ltd.

Wei, Bill, Leila Sauvage, and Jenny Wölk. 2014. Baseline limits for allowable vibrations for objects. In *ICOM committee for conservation 17th triennial meeting*, Paper 1516. Melbourne, Australia: Pulido & Nunes; ICOM Committee for Conservation.

Wei, William. 2017. *Vibration research and testing: What was the question.* Amsterdam: Cultural Heritage Agency of the Netherlands (RCE).

Wei, William, Siobhan Watts, Tracey Seddon, and David Crombie. 2018. Protecting museum collections from vibrations due to construction: Vibration statistics, limits, flexibility and cooperation. *Studies in Conservation* 63: 293–300.

Yahya, Musli N., and Desmond D. V. S. Chin. 2017. A review on the potential of natural fibre for sound absorption application. *IOP Conference Series: Materials Science and Engineering* 226: 012014.

Yang, Sheng, Botao Ma, Jiaqi Ge, Mingbin Li, Ling Zhang, Chen Min, and Jintai Liu. 2020. Overview of preventive conservation research work on seismic mitigation and subway vibration control in Chengdu museum of China. *Studies in Conservation* 65: 212–220.

Zuanni, Chiara, and Campbell Price. 2018. The mystery of the "spinning statue" at Manchester museum. *Material Religion* 14: 235–251.

Measuring Object Deterioration Rates In-Situ

David Thickett

Abstract For optimisation both environmental preventive conservation and modelling of decay require measurement of object deterioration rates. Instrumentation has developed to the point that this is now feasible in some situations with sufficient precautions and understanding of instrument stability and likely errors. Complex environments mean in situ measurements are desirable. The main approaches are described and assessed, with examples to highlight the important issues.

Keywords Object monitoring · Acoustic emission · Pollution · Dimensional change · Mass change · Colorimetry

1 Introduction

Conservation can be defined as the management of change. For preventive conservation, deterioration rates are critical to determine success and to fine tune environments. This is also an essential data set to calibrate any predictive analysis. The complexity of cultural heritage objects has long been recognised. The limitations of surrogates are well documented and only measurements on actual objects, with their complexity and aged nature can provide confidence that modelling approaches deliver a good representation of cultural heritage object response. Also, many, perhaps the majority, of cultural heritage collections reside in environments with fluctuating levels of relative humidity, light, temperature, a wide range of pollutant gases and particulates. Figure 1 shows the relative humidity (RH), temperature, lux, nitrogen dioxide, ozone, sulfur dioxide and particulate concentrations for five size ranges at Apsley House.

This historic house, with a mixed heating regime and pollution filtration in only one room, sits in central London, on an extremely busy traffic roundabout. A period

D. Thickett (✉)
English Heritage Trust, London, UK
e-mail: David.thickett@english-heritage.org.uk

© The Author(s) 2025
Á. F. Perles-Ivars et al. (eds.), *Collection Care*, Springer Proceedings in Archaeology and Heritage, https://doi.org/10.1007/978-3-031-85655-6_10

119

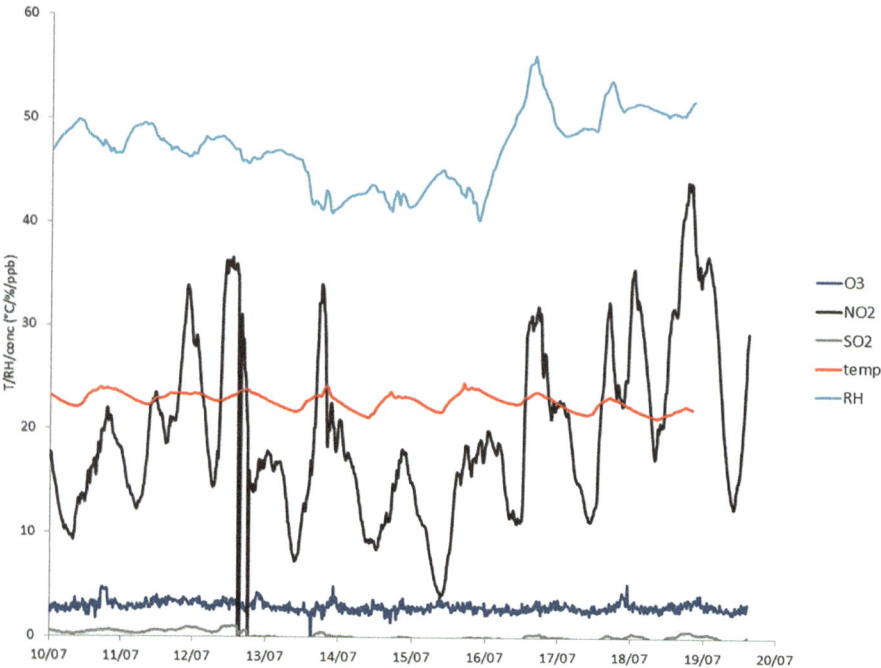

Fig. 1 Environment in Apsley House, London

of fourteen days is shown to clearly show the fluctuating levels, most parameters also show a strong seasonal variation.

This complex situation makes designing representative laboratory experiments to study the deterioration processes, extremely challenging. The environments require many levels, with several different parameters, increasing the work load dramatically. Studying objects in-situ incorporates all the real life complexity. The development of non-invasive and portable instrumentation has allowed this to occur in some instances. There are significant challenges that have had to be overcome and some still to be sufficiently understood. Small experimental details can be critical to successful measurements. Material degradation can be considered as falling under three main categories related to environment; physical, chemical and biological. In cultural heritage, there are frequently synergistic effects between factors, such as deposited dust dramatically increasing metal chemical corrosion rate (Thickett and Costa 2014). There are also interactions between different degradation categories, for example, the chemical acid hydrolysis degradation of cellulose weakening paper, so it breaks at a much lower induced stress from RH fluctuations. Three main types of measurements will be discussed; periodic, continuous and direct tracing. The utility of each method towards the types of damage will be explored.

2 Periodic Measurements

Periodic measurements were the first to be undertaken. Weight can be related to moisture uptake or corrosion. Archaeological iron artefacts displayed in showcases at Battle Abbey increased in mass by 12% over a period of twenty years, despite lacking any obvious signs of active deterioration on their surface. The controlled RH during measurement and display ruled out moisture uptake. The showcases were kept below 40% and measurements planned to occur when the objects had been at 35% for two weeks. Trials showed rapid measurements (within 5 min of opening the showcase did not allow sufficient time for the object mass to increase). This mass increase is related to corrosion, which was not visible on the objects' surfaces.

For many methods, the accurate repositioning of an instrument onto the surface determines the errors and overall sensitivity. Colorimetry is ideally suited to study silver tarnish (Thickett and Hallett 2021). Tarnish can be very localised and small changes in measurement position can affect readings dramatically. Melinex masks, aligned with object decoration, control repositioning for subsequent measurements. Repositioning introduced errors in the order of $\pm$ 0.02 in a* and b* and $\pm$ 0.03 in L* using a Minolta 2600D colorimeter, positioning by eye through the viewport and keeping the colorimeter orientation the same. The colorimeter is calibrated by the manufacturer annually and both white and dark calibrations run between each 10 measurements. Colorimetry has also been used to estimate the climate induced yellowing of varnish in paintings by selecting areas over white paint. At Apsley House very high nitrogen dioxide levels pose a risk to varnishes. The Goya portrait of Wellington from 1812, has a mastic varnish, known to be sensitive to nitrogen dioxide (Odlyha et al. 2007). Whilst monitoring takes place in the room with diffusion tubes, the actual response of an object was measured to evaluate the magnitude of the risk. Periodic monitoring was undertaken with an Ocean Optics 2000 spectrometer with a 1 mm fibre-optic probe. This was thought to present less risk to the paintings as only the 1 mm head of the fibre optic needs to be in contact, not the much larger head of the Minolta colorimeter. Large areas of white pigment were selected to measure over and the relative effect of repositioning the probe manually was assessed, and errors (2 standard deviations) reported accordingly. The measured levels over five years are shown in Table 1.

Table 1 Yellowing of Mastic varnish

Date measured	Delta E_{00}	Concentration nitrogen dioxide (ppb)	Width at half height Mastic (cm^{-1})
Jan 2018	0.32 ± 0.04	5 ± 1	2.230 ± 0.153
Jan 2019	0.25 ± 0.03	3 ± 1	2.345 ± 0.181
Jan 2020	0.28 ± 0.02	6 ± 1	2.634 ± 0.193
Jan 2021	0.68 ± 0.08	12 ± 2	3.876 ± 0.248
Jan 2022	0.71 ± 0.07	16 ± 3	3.981 ± 0.232

As can be seen, the nitrogen dioxide concentrations have increased in the later part of the UKs lockdown due to the pandemic and afterwards. This is correlated with an increase in the apparent yellowing of the mastic varnish. The weakness of this approach is it assumes the white paint under the varnish is not yellowing. It is planned to use 3D microscopy to see if it is possible to determine the relative yellowness of the two layers in situ. To provide more information, mastic samples were placed in the room. These were abraded (1200 grit aluminium oxide fabric) to provide a flat surface which was analysed with FTIR (Bruker Alpha) with a diamond ATR. The assessment was based on the method developed by Odlyha et al. (2007) using width at half height of the carbonyl peak to assess broadening as a measure of oxidation. The values followed the same trend as the mastic yellowing indicated by the colorimetry measurements. The gallery at Apsley has a pollution filtration system which is being recommissioned to reduce nitrogen dioxide levels.

Non- invasive FTIR (Nicolet Inspect IR microscope) has been used to quantify Limoge enamel chemical deterioration over a 15-year period and analyse preventive conservation interventions (Thickett 2020; Thickett et al. 2018). The processes to realign the microscopic analytical areas and long-term stability testing of the instruments were essential to provide useable data of sufficient sensitivity and have been described previously (Thickett et al. 2017, 2018). The monitoring allowed assessment over periods of two years for a series of improvements.

The initial case setup was compromised due to increasing air exchange rate (Thickett et al. 2006). By 2007 the showcase was spending a significant amount of time above 42% RH, the value recommended for unstable glass compositions (Koob et al. 2018). Replacing the buffer material (Artsorb) with one with more capacity (Rhapid gel) improved the environment RH, but the FTIR monitoring proved some of the enamel colours were still showing measureable deterioration in a two-year period. Pollution monitoring indicated the presence of levels of formic acid sufficient to accelerate glass deterioration (over 37 $\mu g/m^3$). The control system was hence changed to a Hahn RK unit, providing both better RH control and pollution control (below detection 21 $\mu g/m^3$). FTIR analyses repeated after 5 and then 8 years showed no detectable deterioration of the enamel surfaces. Results are shown in Table 2.

Periodic measurements, when well-conceived and executed, can work well to track chemical deterioration. They can also provide sufficient evidence for some biological processes. A series of mould damage functions were assessed by comparison with visual assessment (including some under magnification), using some shorter time periods (2–4 weeks) differences in performance could be observed (Thickett et al. 2014). Correlating with environmental parameters can be difficult if not impossible when they vary to a significant extent. The monitoring of physical deterioration is not well suited to periodic methods as the fluctuating RHs generally considered responsible are very difficult to assign over longer periods.

Table 2 FTIR analysis of enamel degradation

Date	Infra-red splitting (cm^{-1})			Percentage time above 42% RH	Intervention/notes
	Purple enamel	Red enamel	Blue enamel		
Sep 2003	65.1 ± 2.3	34.1 ± 2.4	61.0 ± 1.5	4	Showcase AER increased from 0.5 to 2.5d^{-1}
Aug 2007	74.3 ± 1.5	33.5 ± 1.9	65.3 ± 2.1	13	Artsorb replaced with Rhapid gel
Dec 2009	78.7 ± 2.3	35.0 ± 2.4	68.9 ± 1.1	3	Hahn RK2 unit installed
Sep 2014	74.8 ± 3.1	33.8 ± 1.6	66.8 ± 2.3	0	
Sep 2018	74.4 ± 2.5	34.2 ± 2.1	66.9 ± 3.4	0	

3 Continuous Measurements

Continuous measurements can be more easily correlated with high frequency environmental measurements, which is especially important for climate induced deterioration. Several technologies exist for physical measurements, mass or dimensions and were summarised in Thickett et al. (2018) and Dulieu-Barton et al. (2005). Continuous measurements of the mass of five walnut panels, in a showcase, using RDP model 31 load cells running onto model 600 multichannel conditioning system with 611 strain gauge amplifiers and the outputs logged with an Smartreader007 logger. The small size of the load cells allowed their placement between the fittings used the hang the panels, and the fixing hooks. Temperature and RH was measured with a Meaco radiotelemetric transmitter with NAMAS calibrated Rotronic Hygroclip 1 probe.

The measured mass changes are shown in Fig. 2.

The mass changes can be readily related to RH fluctuations. It is impossible to convert these values accurately to moisture contents as the dry mass of the panels is not known and cannot be determined without damage. To determine which mass fluctuations could cause plastic deformation, significant knowledge of isotherms and mechanical properties is required. Other methods have been used and were reviewed in Thickett et al. (2018). The deformetic kit has been developed for panel paintings, but requires rigid attachment of fittings to the reverse of the panels and space behind the panel. Non-invasive digital image correlation is finding wider application in cultural heritage. The surface needs sufficient detail and for many objects, such as portrait paintings, this poses limitations (Thickett et al. 2018). All these methods similarly require interpretation of the measured strain or mass to determine when reversible changes progress to irreversible ones. This is significantly complicated by hysteresis, viscoelastic behaviour, creep and fatigue.

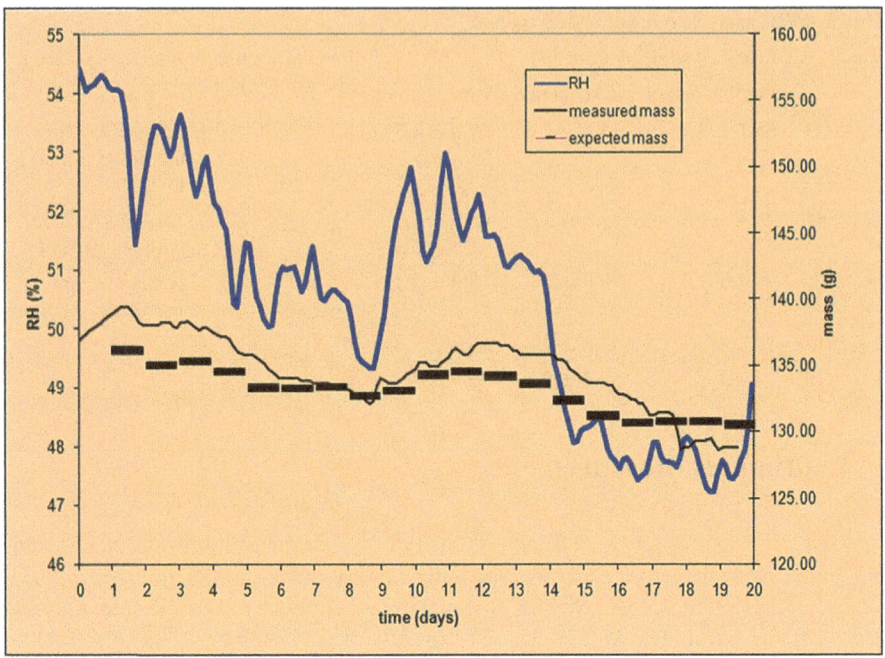

Fig. 2 Mass change of walnut panels

4 Direct Tracing

One method is available that produces signals more closely aligned with damage. Acoustic emission has been developed over the past decade to monitor furniture, enamels, metal corrosion and salt activity and damage in stone. It is widely used in engineering, but sensor attachment, data interpretation and filtering of background noise all have to be tailored for cultural heritage applications. The technique can be extremely sensitive, with this mainly limited by the background noise at the monitoring location. This especially affects furniture monitoring. Wood response is relatively weak (Kawamoto and Williams 2002). Fracture experiments measuring increasing crack length and hence increase in surface area against acoustic energy released showed the following response factors; high lead enamel 300 k/mm^2, ivory 231 k/mm^2, Caenstone 430 k/mm^2, walnut (equilibrated at 33% RH) 17 k/mm^2, walnut (equilibrated at 54% RH) 12 k/mm^2. The acoustic emission produced on fracturing, which is usually centred around 150 kHz is strongly absorbed by the wood and affected by growth direction and moisture content.

Several methods have been developed to filter out, or identify, noise. Most researchers use some form of filtering of the frequency of signals and a background level above which signals are not recorded. Anti-correlation is frequently used to monitor wooden objects (Strojecki et al. 2014). Two sensors are placed sufficiently apart, so they do not detect signals originating from the other sensor. Reported detection distances in wood are 6–20 cm. This is most frequently checked with a Hsu Nielsen device, essentially a 2B pencil with a holder to help control the lead break. Some researchers have questioned the repeatability of this approach (Sause 2011). The method has been improved using a Physical Acoustics Pocket AE with WD sensors to generate narrow acoustic emission pulses centred on 150 kHz, to mimic the signal from micro-fracturing wood. The sensor is tracked over the surface to produce a map of the detection limit for the sensors applied to the piece of furniture. In anti-correlation, a signal that is detected by both sensors simultaneously is determined to be noise as the signal should not travel between the sensors. A signal detected on one sensor only is considered acoustic emission from the wood. An alternative approach is to use multiple sensors, at least three and only accept signals that originate from between the sensors. The signal lag maximum is set by experimenting with the Pocket AE signal generator.

Two acoustic emission sensors were pushed against a nineteenth century card table, with Melinex between the sensors and the table. These were Physical acoustics WD sensors, running onto a Physical Acoustics Micro II system. The table in its historic context at Apsley House is shown in Fig. 3. In historic interiors the monitoring cannot be too visually or intrusive in other ways, noise, vibration and electricity sockets are often limited. The system accommodates long sensor to mains distances, and allowed the main parts to be placed under the adjacent sofa, limiting the visual impact. The room utilizes a floor heat matt controlled on humidistatic principles via a BMS using data from the Meaco monitoring system. Issues with the control logic were observed, with RH periods below 40% occurring. The table was monitored for 30 months. Twelve months of initial data is shown in Fig. 3. The acoustic energy observed is also included in Fig. 3.

As can be seen, acoustic emission is registered rapidly after some points when the RH drops below 40%, but not others. These levels of acoustic emission energy relate to extremely small cracks, each event equates to less than 20 μm^2. For this wood thickness, this would generate a total crack length of less than 1 mm in 50 years. The BMS was reprogrammed after 12 months, with an increased RH set-point to turn the heating matt off. This reduced the low RH periods, but did not remove them and acoustic emission was still being registered, hence the RH set point was increased again. No acoustic emission was registered in the following 12 months.

The acoustic emission and RH data from monitoring 8 pieces of furniture in 7 sites was collated. Three published damage functions were applied to the RH data to predict damaging periods. The damage functions used were;

- percentage dimensional change—from Image Permanence Institute, based on United States Forestry Product Laboratory isotherm data, with damage assumed when index is greater than 1.5% (Nishimura 2007)
- Climate Tool Box—Boris Pretzel (Pretzel 2014)
- HERI-e,—Jerzy Haber Institute of Catalysis and Surface Chemistry, herie.pl, using a strain level of 0.005 (Kupczak et al. 2018).

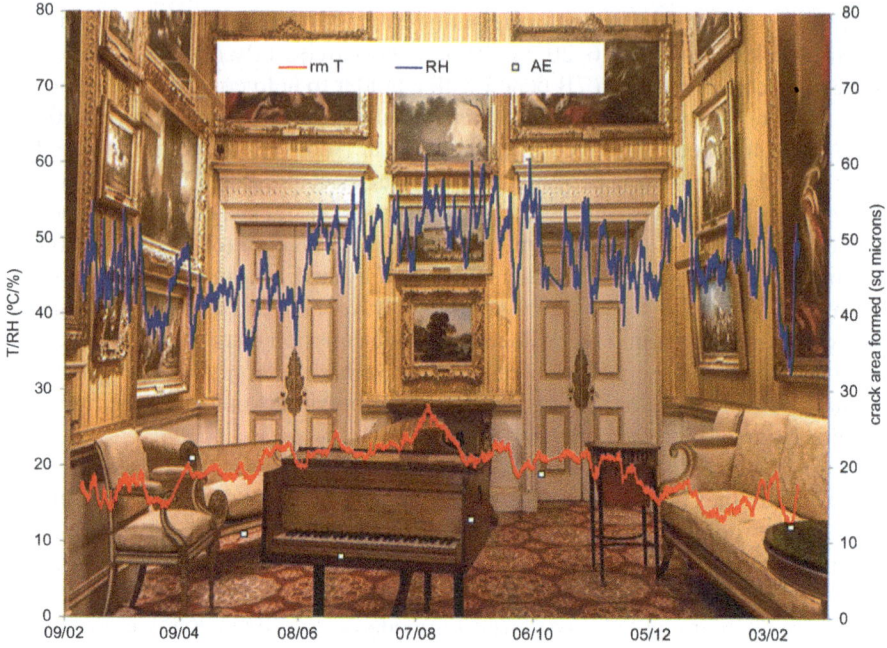

Fig. 3 Climate and acoustic emission measured

The Climate for Culture damage function was not investigated as this does not allow discrimination of the actual time period predicting damage without extensive data processing. The number of damaging events predicted, were compared to the acoustic emission events monitoring. The results were expressed as a simple percentage agreement and are shown in Table 2.

For the card table IPI predicted 67% of the events observed, climate tool box performed better, predicting 90% and Heri-e performed best at 96%. The collated results for the 8 pieces of furniture are included in Table 3. The HERI-e predictions best matched the observed data. The IPI function appears to be under-sensitive for this material. Clearly a larger body of results is required to draw more general conclusions. Measurements are continuing at English Heritage and a generous investment from United Kingdom Research Council has quadrupled capacity for such experiments.

5   Conclusions

Under some circumstances it is possible to analytically measure object deterioration rates. This relies on good long-term instrument stability, careful calibration and reliable methods to repositioning analysis or connecting sensors to object surfaces. It can determine much smaller changes than those visible to the naked eye or with

Table 3 Correlation of three damage functions with acoustic emission measurements

Object	Predicted AE damage instances (% of those measured)		
	IPI	Climate tool box	HERI-e
Apsley, yellow striped room, card table	67	90	96
Walmer, Wellington bedroom, wash table	73	92	97
Walmer, Pitt room, chest of drawers	75	93	97
Average of 8 pieces of furniture			

Damage function	Predicted AE damage instances (% of those measured)	Predicted instances when no AE observed (%)
IPI	66	0
Climate tool box	84	12
HERI-e	92	3

accessible portable magnification. It can also detect types of deterioration that lead to no visible changes, or no changes in the earlier stages, such as oxidation and acid hydrolysis. Periodic measurements are reasonable for chemical deterioration and can give information on biological deterioration. The complex RH fluctuations frequently observed in cultural heritage environments, make assigning observed physical changes very difficult and often impossible. Continuous measurement can allow correlation with RH fluctuations, but the methods presently available (mass and dimensional change) need significant additional information to determine if the observed changes are reversible. Acoustic emission gives a more direct measure and signals can be correlated to the area of new cracks produced. Background noise is a limiting factor, especially in situations with public access and complex shock, vibration and electromagnetic environments (Diodati 2001). Anti-correlation or signal location through multiple sensors, provide methods that work to filter out the background noise. Acoustic emission has been demonstrated to be very effective in improving climate management. Correlation of acoustic emission from a small corpus of furniture measurements and three damage functions using simultaneous RH data has indicated differences in performance between the damage functions.

References

Diodati, P., S. Piazza, A. Del Sole, and L. Masciovecchio. 2001. Daily and annual electromagnetic noise variation and acoustic emission revealed on the Gran Sasso mountain. *Earth and Planetary Science Letters* 184: 719–724.

Dulieu-Barton, Janice M., Leonidas Dokos, Dinah Eastop, Frances Lennard, Alan R. Chambers, and Sahin Melin. 2005. Deformation and strain measurement techniques for the inspection of damage in works of art. *Studies in Conservation* 50: 63–73.

Kawamoto, Sumire, and Sam R. Williams. 2002. *Acoustic emission and acousto-ultrasonic techniques for wood and wood-based composites.* United States Department of Agriculture Forest Service Report, FPL-GTR-134.

Koob, Stephen P., N. Astrid, R. Van Giffen, Jerzy J. Kunicki-Goldfinger, and Robert H. Brill. 2018. Caring for glass collections: The importance of maintaining environmental controls. *Studies in Conservation* 63: 146–150.

Kupczak, Arkadiusz, Mariusz Jędrychowski, Marcin Strojecki, Leszek Krzemień, Łukasz Bratasz, Michał Łukomski, and Roman Kozłowski. 2018. HERIe: A web-based decision-supporting tool for assessing risk of physical damage using various failure criteria. *Studies in Conservation* 63: 151–155.

Nishimura, Dwight. 2007. *Understanding preservation metrics.* Rochester, NY: Image Permanence Institute.

Odlyha, Marianne, Charis Theodorakopoulos, David Thickett, Morten Ryhl-Svendsen, J.M. Slater, and Roberto Campana. 2007. Dosimeters for indoor microclimate monitoring for cultural heritage. In *Museum microclimates*, eds. Padfield Tim, and Borchersen Keto, 73–79, Hvidovre, Denmark: LP Nielsen Bogtryk.

Pretzel, Boris. 2014. Reasonable—broadening acceptable climate parameters for furniture on open display. In *Proceedings of the ICOM-CC 17th triennial conference preprints*, ed. Bridgland Jack, 15–19, Melbourne, Australia: International Council of Museums.

Sause, Markus GR.. 2011. Investigation of pencil-lead breaks as acoustic emission sources. *Journal of Acoustic Emission* 29: 184–196.

Strojecki, Marcin, Micha Lukomski, Leszek Krzemie, Joanna Sobczyk, and Lukasz Bratasz. 2014. Acoustic emission monitoring of an eighteenth-century wardrobe to support a strategy for indoor climate management. *Studies in Conservation* 59: 225–232.

Thickett, David. 2020. Review of analysis for cultural heritage conservation. *Current Topics in Analytical Chemistry* 12: 73–88.

Thickett, David, Cho S. Cheung, Hong Liang, John Twydle, Greyson R. Maev, and Denis Gavrilov. 2017. Using non-invasive non-destructive techniques to monitor cultural heritage objects. *Insight-Non-Destructive Testing and Condition Monitoring* 59: 230–234.

Thickett, David, and Virginia Costa. 2014. The effect of particulate pollution on the corrosion of metals in heritage locations. In *ICOM-CC 17th triennial conference preprints*, ed. Bridgland Jack, 15–19. Melbourne, Australia: International Council of Museum.

Thickett, David, Frances David, and Naomi Luxford. 2006. Air exchange rate: A dominant parameter for showcases. *The Conservator* 29: 19–34.

Thickett, David, and Kenny Hallett. 2021. Managing silver tarnish. In Transcending boundaries: Integrated approaches to conservation. In *ICOM-CC 19th triennial conference preprints*, ed. Bridgland Jack, Beijing, China: International Council of Museum.

Thickett, David, Paul Lankester, and Pereira L. Pardo. 2014. Testing damage functions for mould growth. In *Proceedings of the ICOM-CC 17th triennial conference preprints*, ed. Bridgland Jack, 15–19, Melbourne, Australia: International Council of Museum.

Thickett, David, Vulkaner Vilde, Phoebe Lankester, and Eimear Richardsen. 2018. Using science to assess and predict object response in historic house environments. Postprints of preventive conservation in historic houses and palace museums: Assessment methodologies and applications. Milan, Italy: Silvana Editoriale.

A Study of Three Modern Asian Lacquers Using Surface Metrology and Data Science/Analytics

H. David Sheets, Patrick Ravines, and Marianne Webb

Abstract This paper presents a quantitative approach to the study of surfaces using surface metrology and data science techniques. Asian lacquer types with various additives have a quantifiable impact on the topography of lacquered surfaces that may be used to detect lacquer type from non-contact measurements. To understand the unaged and aged characteristics, 15 different formulas of Asian lacquer were prepared using laccol, thitsi, and urushi with a range of oils, pigments, and resins were examined. The surfaces of the Asian lacquers test specimens were studied using confocal microscopy to acquire quantitative surface texture data, and data science methods of feature engineering and convolutional neural networks (CNN) were applied to analyze the numerical surface texture data, and assign lacquer specimens to the three lacquer types. Correct classification rates reached as high as 96%.

Keywords Asian Lacquers · Laccol · Thitsi · Urushi · Confocal microscopy

1 Introduction

Surface metrology has been around for decades, has grown considerably and has been introduced to the world of conservation and preservation. The surfaces of works of art and cultural heritage objects are equally affected by various physico-chemical processes, including heat and cooling, absorption and desorption of liquids and gases, electrical charges, abrasion, and light-induced changes. We present here the results

H. David Sheets (✉)
Data Analytics Program, Canisius College, Buffalo, NY, USA
e-mail: sheetsh@merrimack.edu

P. Ravines
Patricia H. and Richard E. Garman Art Conservation Department, Buffalo State College, Buffalo, NY, USA
e-mail: ravinepc@buffalostate.edu

M. Webb
Webb Conservation Services, Vancouver, BC, Canada
e-mail: mw@mariannewebb.com

© The Author(s) 2025
Á. F. Perles-Ivars et al. (eds.), *Collection Care*, Springer Proceedings in Archaeology and Heritage, https://doi.org/10.1007/978-3-031-85655-6_11

of a pilot study on classifying surfaces coated with Asian lacquers to specify lacquer type based on non-contact optical measures using feature engineering (the roughness spectra) and machine learning methods (convolutional neural networks). Effective classification of surfaces to lacquer type was possible despite different lacquer additives and differences in surface aging.

A valuable property of light-based surface metrology techniques is that they are non-perturbing (non-invasive, non-contact, and non-destructive) methods for the surface examination of original historic and artistic works. Published studies address a variety of objects such as ancient coinage (Grynszpan et al. 2004), silver gelatin photographs and daguerreotypes (Ravines et al. 2008, 2010), consolidation of porous archaeological objects with cyclododecane (Peters et al. 2018), and the surface of paintings on canvas and icons (Elias et al. 2006; Sotiropoulou and Wei 2006; Delaney et al. 2008; Childers 2019). Surface metrology techniques can be independent of illumination and reflectance factors and important in the study of delicate and sensitive surfaces.

This paper focuses on the application of a surface metrology instrumental technique of confocal microscopy, as developed in 1955 by Minsky (1988), to study Asian lacquer surfaces. Confocal microscopy is a technique that offers the ability to quantitatively image surface textures, thereby allowing one to visually experience the 3D nature of small-scale features and to create digital surface scans for subsequent analysis.

Laccol, thitsi, and urushi were the three Asian lacquers used to prepare modern test panels using traditional recipes with various additives, Fig. 1. The characterization of the surfaces of Asian lacquer panels is a portion of a broader research initiative at the Getty Conservation Institute Web Site (2010). To study the aging characteristics and develop cleaning methods for Asian lacquers, 15 different formulas of Asian lacquers using laccol, thitsi and urushi were prepared in 2017 with the most common additives: oils, pigments, and natural resins. The urushi and laccol lacquers were purchased commercially from Japan and Taiwan. Suppliers did not provide composition information, and both were used as purchased. Unfiltered thitsi was sourced from the Department of Forestry, Ngao District, Lampang Province, Thailand. The formulas within each of the three lacquer categories differ by one ingredient from the next in the series. The preparation of the panels followed traditional recipes and protocols to minimize differences and ensure standardization of the final products.

Data science tactics were used on the confocal microscope surface scans to determine if it was possible to determine the lacquer type on a surface based on the surface topography. Feature engineering was used to derive a "roughness spectrum" for the surfaces, which could then be examined using Principal Component Analysis (PCA) and classified to lacquer using a Random Forest classifier. Additionally, a convolutional neural network classifier, similar to those used in image identification, was created to classify the surfaces.

Included in this study are two lacquer objects from the Garman Art Conservation Department study collection made by unknown artists/craftsmen in the past 20 years in South East Asia. Art historical material information on the two lacquer objects lists the pillbox as thitsi-based and the plate as laccol-based. The thitsi box was

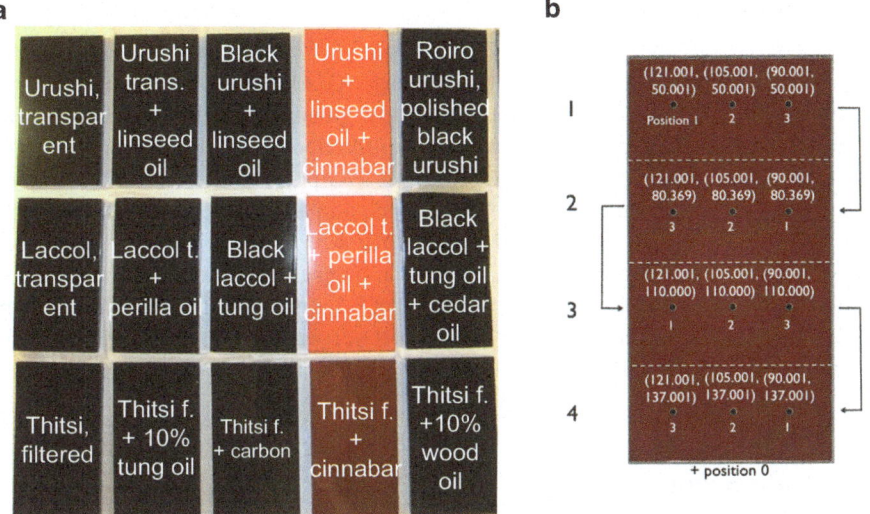

Fig. 1 **a** The 15 lacquer panels prepared with laccol, thitsi and urushi and additives for this study. One thitsi plate was damaged so 14 plates were studied. **b** A sample Asian lacquer cinnabar plate depicting the 13 regions of interest measured and showing the (x, y) coordinates of each for repeatability and future use. 14 panels were measured this way with a 10 × objective

verified by pyrolysis gas chromatography as lacquered with thitsi. While the second art object was historically identified as a laccol plate, chromatography indicated it was not a lacquer finish but a modern commercial coating. The two objects were added to test the premise of using surface textures and artificial intelligence to identify, characterize and classify lacquer objects. Inclusion of the falsely labeled laccol plate in the study provides an interesting control, as it should not be classified as any of the three lacquers in the study.

2 Experimental Design

The contemporary Asian lacquer panels studied were made of laccol, thitsi, and urushi, both pure and with various traditional additives, Fig. 1a. Marianne Webb prepared the laccol and thitsi panels in her studio in Vancouver, British Columbia, Canada, and Sunhwa Kim made the urushi panels in Buffalo, New York, USA, in 2017. Aging of segments of each panel was carried out for 100, 200, 300, and 400 h with a Xenon lamp exposure at 340 nm and submitted to relative humidity cycling of 80% per week followed by 20% for eight weeks, Fig. 1b.

2.1 Confocal Microscope

A μsoft 3D measurement system (NanoFocus AG, Oberhausen, Germany) based on the confocal scanning disk principle with a green light source was used with a $10 \times$ objective (https://www.nanofocus.com/products/usurf/usurf-custom/). The confocal system provides data arrays of 984×984 points/pixels for all objectives. The surface characterization of the 14 Asian lacquer panels (one panel was damaged and unusable) was done using a $10 \times$ objective with a numerical aperture of 0.3, working distance of 11.0 mm, 1600×1600 mm^2 measurement field, and with vertical resolution of 20 nm and lateral (xy) resolution of 3.1 μm and an $n = 111$ digital scans used in the study. Commercial software (μsoft software version 6.1) was used for the operation of the confocal microscope and data acquisition.

2.2 Data Science

Feature engineering processes extract useful derived variables from a data set that act as features (summary variables) for further analysis. In ordinary life, people use a range of features to describe objects in the world around them, and although feature engineering is mathematical in nature, it is akin to this normal human process. In the typical approach to estimating waviness and roughness of a surface, a spatial filter (with a size set to 10% of the surface area) is applied to the data to smooth it and yield an estimated waviness (the smoothed surface) (Raja et al. 2002). The root-mean-square (RMS) of the height value, Sq, is taken as an overall measure of roughness of the surface. More complex approaches to surface characterizations have been developed, including those based on discrete Fourier transforms (Massaro et al. 2020) or other advanced feature engineering methods (Vorburger et al. 2016). Within this study, the Gaussian filter process has simply been extended. A series of filters of increasing size has been used to compute a roughness and a waviness/ contour for each size of the filter, rather than apply a single filter of fixed size to produce a single roughness and smoothed/waviness surface. The smallest filter size is taken as the resolution limit of the measurement, meaning a single pixel. The filter size is steadily increased at some desired rate so that the residuals of the filter produce a roughness estimate at the scale of each filter used. The RMS/Sq value of the residuals at each filter size forms a roughness spectrum, describing the magnitude of the roughness at a variety of scales. The entire roughness spectrum of a sample forms a set of engineered features called a feature vector (Vorburger et al. 2016) that describes the texture of the surface as a series of scale-dependent measurements rather than a single RMS/Sq roughness. This iterative filtering process was carried out in R (R Core Team 2020) using a robust Gaussian filter from the spatstat package. The resulting roughness spectra may then be analyzed with a wide range of multivariate statistical methods. Principal Component Analysis is used to examine the broad patterns of variance in the data, and a Random Forest Classifier is used to test the ability to assign a roughness spectrum to a specific lacquer.

The second approach uses a deep learning method called a convolutional neural network (CNN) (Chollet 2018), which has proven effective in image analysis in detecting and identifying objects in an image. The surface scans are treated as monochrome images, in which the height at each location is false-color coded in shades of grey. The CNN employed N x N pixel patches of the entire image so that many different training examples were available from each of the 984×984 data scans available. Patch sizes of 50×50, 100×100, 150×150, and 200×200 pixels were examined to determine the effect of patch size on the analysis. The CNN was implemented in Python using the Keras API to the TensorFlow library (Chollet 2015). A simple CNN structure of 3 paired convolution and pooling layers, followed by a flattening layer, one hidden layer for the classification, and an output layer was created, trained, and evaluated, based on examples from Chollet (2018). A Dropout Bayesian method (Dürr and Sick 2020) was used to assess the quality of assignments so that specimens with uncertain source affinity could be detected and not assigned falsely to any specific lacquer type.

3 Results and Discussion

The roughness spectra were calculated for all surface scans from all samples, both the aged and non-aged of laccol, thitsi, and urushi lacquers with additives, using a total of 19 different filter sizes ranging from 1.62 microns up to 162 microns in size (1 to 200 pixels). The mean and standard deviation were then calculated for each lacquer type and the two art objects, based on the non-aged surfaces, as shown in Fig. 2. There are apparent differences in the mean values of the roughness spectra for the three lacquers. The thitsi box has a pattern generally consistent with the thitsi specimens, while the laccol plate has marked differences in the roughness spectra, notably at larger spatial scales.

As each individual specimen is represented by the 19 values in the roughness spectra, an ordination method called Principal Components Analysis (PCA) (Pearson 1901; Vorburger et al. 2016) was used to produce a reduced dimensionality plot, Fig. 3. PCA plots require practice to interpret but provide a simplified depiction of the significant patterns of variation in the data. If different groups of specimens (such as lacquer types) group or cluster on the diagram, it indicates that specimens with similar group variables (lacquer type or age) also have similar patterns in the roughness spectra. The PCA does not use lacquer type information to estimate the axes. PCA is a form of unsupervised learning that reveals patterns in data without source or outcome labels.

The first two PCA axes based on the analysis of the roughness spectra of 111 specimens in our study show a clear cluster or segregation of specimens by lacquer type. The aged urushi specimens do separate substantially from the unaged specimens (age coding is not shown in Fig. 3 for reasons of clarity). Aged specimens of the other

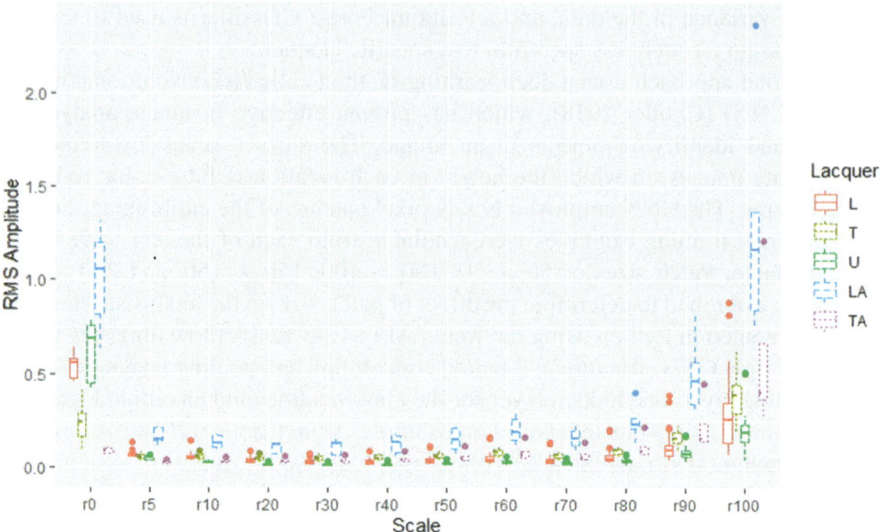

Fig. 2 The roughness spectra of the three lacquer types and the two art objects shown as a box plot. The x-axis range indicates the filter size in pixels, with R0 indicating the residuals of the 5-pixel filter R5. Each pixel is 1.62 microns. The laccol plate (LA), in blue, shows higher roughness at virtually all scales. The thitsi box (TA), in purple, has mean values consistent with the thitsi lacquer specimens

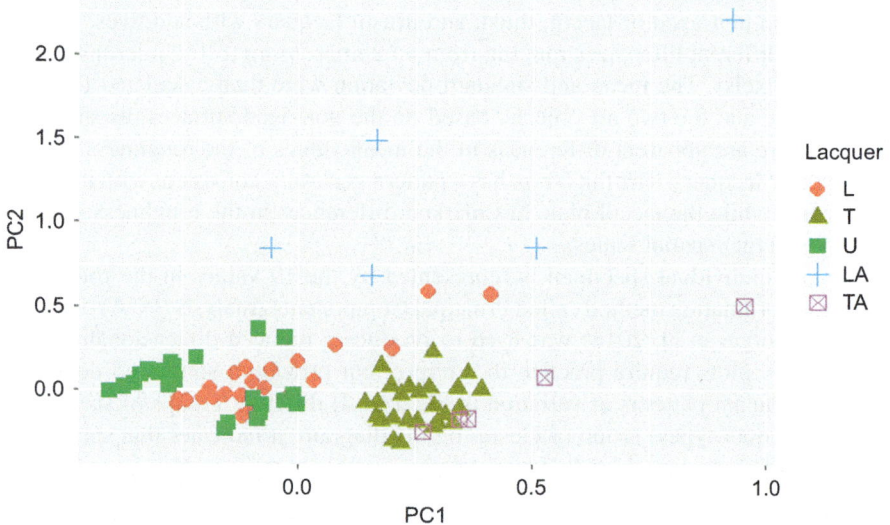

Fig. 3 A PCA (or ordination) plot showing the major patterns of variance in the roughness spectra. The segregation of the points (representing specimens) on this plot indicates a grouping of specimens by lacquer type, indicating that the dominant pattern of variation among specimens is strongly associated with the lacquer type. The two art objects (LA, TA) are also shown

Table 1 Rates of assignments of lacquer samples to lacquer type by the Random Forest (RF) classifier using the roughness spectra, and by the CNN method. The number of scans used (n) was 70 for unaged specimens, 41 aged specimens and 5 each for the two art objects

Group	Correct (%)	Incorrect (%)	Inconclusive (%)
Unaged, RF	96	4	
Aged, RF	86	14	
Thitsi box	100	0	
Laccol plate	40	60	
Unaged, CNN	94	0	6
Aged, CNN	95	0	5
Thitsi box	100	0	0
Laccol plate	20	20	60

two lacquers remain close to the unaged specimens. The thitsi box appears close to the thitsi specimens, but the laccol plate does not plot particularly close to any of the lacquer specimens.

The roughness spectrum can also be used in a supervised learning method to classify specimens to a lacquer type. In this case, a random forest method using 500 decision trees was trained to classify specimens into lacquer types, utilizing the roughness spectra as the input data. The RF classifier was fitted (trained) with the unaged lacquer specimens only, and the performance was assessed using the unaged and aged lacquer specimens and the two art objects, Table 1. Not surprisingly, the RF classifier performed best on the unaged data it was trained on, with the classification rate falling by ten percentage points for the aged specimens. The thitsi box was always classified as thitsi, while the laccol plate assignments were no better than random.

The CNN-based classifier rapidly reached high levels of correct classification at all patch sizes, with the largest patches, Table 1, achieving the highest rates of correct assignment of the aged and unaged lacquer specimens. The use of the Dropout Bayesian criteria prevented all false assignments, successfully identifying these as poor-quality assignments and moving them to the inconclusive category instead of the correct or incorrect categories. Again, the thitsi box was always correctly identified, while the laccol plate was most often classified as inconclusive.

4   Summary/Conclusions

The combination of surface metrology and data science has been shown to be a potentially useful tool to quantitatively describe the surfaces of the laccol, thitsi and urushi test panels, achieving high levels of successful assignment of surface to a lacquer type despite differences in lacquer composition and aging. The thitsi pillbox nestled in with the thitsi data on the roughness spectra plot (Fig. 2) and the PCA plot (Fig. 3), whereas the laccol plate did not and is thus suspect. Assignments based on both the roughness spectra and the CNN were entirely consistent for the thitsi box and very poor for the laccol plate. The CNN dropout process was helpful here in indicating that the laccol plate was often inconclusive, meaning difficult to assign to any lacquer category.

The feature engineering-based approach led to the development of a roughness spectra that produces scale-dependent characterizations of the roughness that can be used in subsequent analyses. A simple plot of the mean and standard deviations of the roughness spectra for the three lacquers in unaged form shows clear differences in the roughness spectra (Fig. 2), most noticeable at the smallest scales (R0, R5, R10). A PCA analysis of the data shows clear segregation of the three unaged lacquers, with aged laccol and thitsi specimens remaining close to the unaged specimens, while urushi specimens changed in more complex ways. It should be noted that one person prepared the thitsi and laccol panels, while another person prepared the urushi, so the source of the variation cannot entirely be determined. Clearly, though, there were detectable changes in the roughness spectra associated with specimen aging. Effective classification of aged specimens based on the roughness spectra will require a detailed understanding of the specimen changes related to aging and larger sample sizes.

A random forest classifier trained on unaged specimens performed well in classifying unaged specimens, but the performance deteriorated when aged specimens were classified with the same random forest classifier. The CNN classifier made use of the raw data matrix without any feature engineering and showed better performance than the random forest classifier operating on the roughness spectra. The use of the Dropout Bayesian approach allowed effective identification of poor-quality assignments in the CNN, eliminating all false assignments in this data set. Given more data, it will be possible to train both classifiers on both aged and unaged data, thus incorporating information about the aging process into the classifier to improve performance.

This study shows that advanced analytic approaches can be effective in extracting information from surface texture scans. Both feature engineering and convolutional neural network-based approaches can successfully classify surfaces to the correct lacquer types, as well as identifying and quantifying the impact of surface aging.

Acknowledgements The authors gratefully acknowledge Sunhwa Kim, associate professor, Art and Design Department, Buffalo State College, for preparation of the Urushi lacquer panels, Rebecca Ploeger for pyrolysis gas chromatograph mass spectrometry data on the two lacquer objects: the laccol plate and the thitsi pillbox; and Getty Conservation Institute colleagues Michael Schilling and Herant Khanjian.

References

Childers, Rachel. 2019. *Analysis of varnish effects on blanching using confocal microscopy*, Garman Art Conservation Department, Buffalo State College.
Chollet, François. 2015. Keras. https://github.com/fchollet/keras. Accessed 26 June 2020.
Chollet, François. 2018. *Deep learning in python*. Shelter Island, NY: Manning Publications.

Delaney, John K., E. René de La Rie, Mady Elias, Li Piin Sung, and Kathryn M. Morales. 2008. The role of varnishes in modifying light reflection from rough paint surfaces. A study of changes in light scattering caused by variations in varnish topography and development of a drying model. *Studies in Conservation* 53: 170–186.

Dürr, Oliver, and Beatte Sick. 2020. *Probabilistic deep learning: With python, keras and tensorflow probability*. Shelter Island, NY: Manning Publications.

Elias, Mady, E. René de La Rie, John K. Delaney, Eric Charron, and Kathryn M. Morales. 2006. Modification of the surface state of rough substrates by two different varnishes and influence on the reflected light. *Optics Communications* 266: 586–591.

Getty Conservation Institute Web Site. 2010. Characterization of European and Asian lacquers. www.getty.edu/conservation/our_projects/science/lacquers/. Accessed 20 Dec 2020.

Grynszpan, R.I., J.L. Pastol, S. Lesko, E. Paris, and C. Raepsaet. 2004. Surface topology investigation for ancient coinage assessment using optical interferometry. *Applied Physics A* 79: 273–276.

Massaro, Alessandro, Giovanni Dipierro, Emanuele Cannella, and Angelo Maurizio Galiano. 2020. Comparative analysis among discrete Fourier transform, K-means and artificial neural networks image processing techniques oriented on quality control of assembled tires. *Information* 11: 257.

Minsky, Marvin. 1988. Memoir on inventing the confocal scanning microscope. *Scanning* 10: 128–138.

Pearson, Karl. 1901. On lines and planes of closest fit to systems of points in space. *The London, Edinburgh, and Dublin Philosophical Magazine and Journal of Science* 2: 559–572.

Peters, N., Aaron Shugar, Lucy Skinner, Rebecca Ploeger, and Patrick Ravines. 2018. Microscopic observations to track the behaviour of cyclododecane crystallisation and the effect of crystal formation on fragile porous substrates. In *Subliming surfaces: Volatile binding media in heritage conservation*, ed. C. Rozeik, 67–68. Cambridge: University of Cambridge Museums.

R Core Team. 2020. R: A language and environment for statistical computing. https://www.r-project.org/. Accessed July 2019.

Raja, J., B. Muralikrishnan, and Shengyu Fu. 2002. Recent advances in separation of roughness, waviness and form. *Precision Engineering* 26: 222–235.

Ravines, Patrick, Jiuan Jiuan Chen, and Christian M. Wichern. 2010. Surface characterization and monitoring of surface changes after conservation treatments of silver gelatin photographic papers using confocal microscopy. *Scanning* 32: 122–133.

Ravines, Patrick, Ralph Wiegandt, Richard Hailstone, and Grant Romer. 2008. Optical and surface metrology applied to daguerreotypes. In *Conservation science 2007: Papers from the Conference held in Milan, Italy 24–26 May 2007*. London: Archetype Publications.

Sotiropoulou, S., and William Wei. 2006. Non-contact method for the documentation, evaluation and monitoring of conservation treatments for icons. In *ICOM-CC international meeting icons: approaches to research, conservation and ethical issues*. Athens, Greece: New Benaki Museum.

Vorburger, T.V., J. Song, and N. Petraco. 2016. Topography measurements and applications in ballistics and tool mark identifications. *Surface Topography: Metrology and Properties* 4: 013002.

140   H. David Sheets et al.

Risk Management

Hygro-Mechanical Modelling of Wooden Cultural Objects for Damage and Risk Assessment as a Tool for Preventive Conservation

Josef Stöcklein, Gerald Grajcarek, Michael Mäder, Silvia Oertel,
Andreas Schulze, and Michael Kaliske

Abstract The description at hand presents an overview of the Cultwood research project at the Technische Universität Dresden, Institute for Structural Analysis, and introduces some of the results obtained so far. In the validation project, a tool for damage and risk assessment of conservation and restoration measures on wooden art objects is developed using the finite element simulation method. The tool is applied to two historical art objects. In addition to the development of the complex material models, the results of preliminary experiments carried out to identify parameters and to validate the interaction behaviour of wood and paint layers are examined. Within this publication, the used materials and the experimental and numerical methods as well as selected experimental and numerical results are presented. The advantages of the tool for the restoration of wooden artwork and the application range are discussed.

Keywords Numerical simulation · Hygro-mechanical behaviour · Parameter identification · Validation experiments · Replicas

1 Introduction

Preventive conservation measures as well as conservation and restoration treatments carried out for museum collections are usually chosen according to available options and on the basis of the current state of knowledge or recommendations. In recurring situations of collection care such as art logistics, inappropriate storage conditions or

J. Stöcklein · G. Grajcarek · M. Kaliske (✉)
Institute for Structural Analysis, TUD Dresden University of Technology, Dresden, Germany
e-mail: michael.kaliske@tu-dresden.de

M. Mäder
Dresden State Art Collections, Dresden, Germany

S. Oertel · A. Schulze
Art Technology, Preservation and Restoration of Artistic and Cultural Assets, Dresden University of Fine Arts, Dresden, Germany

© The Author(s) 2025

Á. F. Perles-Ivars et al. (eds.), *Collection Care*, Springer Proceedings in Archaeology and Heritage, https://doi.org/10.1007/978-3-031-85655-6_12

new exhibition presentations, immediate or long-term damages due to the changing environmental influence onto the object can often only be vaguely estimated. The current progress in numerical modelling and analysis allows to support the conservation measures by numerically simulated predictions of the material response and the structural behaviour for different climatic scenarios. The influence of climatic conditions, restoration measures and storage conditions on the damage of wooden cultural objects can be estimated and, thus, qualitative statements about the suitability of treatments and measurements can be determined by a simulation tool for risk management.

This paper provides an overview on some results of the research project Cultwood[1] at Technische Universität Dresden, Institute for Structural Analysis, in collaboration with the Dresden University of Fine Arts (HfBK) and the Dresden State Art Collection (SKD). In this validation project, complex hygro-mechanical finite element material models are developed and validated using replicas of two historical museum objects. Therefore, a panel painting by Lucas Cranach the Elder (1506) of the altarpiece of St. Catherine from the castle church in Wittenberg (today collection of SKD) and a painted cupboard from the eighteenth century with Upper Lusatian provenance (today collection of SKD) are chosen as benchmark examples. In a first step, experiments are conducted to investigate the material behaviour of the used wood species and painting materials and to identify model parameters for the numerical simulation. In a further step, experiments for the validation of moisture induced stresses and the interaction of wood and coatings are realised. To validate the simulation of the behaviour of complex structures, the developed replicas are stored in two climate scenarios and the deformation is measured by photogrammetry.

The aim of the project mentioned is to evaluate the capabilities of this analysis tool on the basis of the validation experiments, to show the advantages and limits of the numerical modelling and to develop necessary research tasks for the further development and specialisation of this tool. Furthermore, the aim exists to disseminate this method and analysis tool in conservation and restoration as well as in heritage conservation sciences.

2   Materials and Methods

2.1   General

For the specification and development of the material models for altogether six replicas, extensive preliminary tests are carried out. These are categorised in three groups and served as parameter experiments to determine necessary material properties, combined experiments split in parameter determination and validation for the

---

[1] Project 03VP05781, funded by the Federal Ministry of Education and Research: Numerical analysis tool for the simulation and risk assessment of climatic and mechanical loaded wooden cultural artwork—Cultwood.

material parameters itself and pure validation experiments for investigation of the interaction between different materials.

The experiments for pure material parameter identification involve tests on the swelling and shrinkage behaviour[2] of the wood species (*Pinus spp., Tilia spp., Picea spp.*), which are macroscopically determined at the original art objects, and tests to determine the diffusion properties of different paint layers. The latter are characterised based on the original object, too, and are associated with their binder systems and colorants. Altogether, eight different paint layers (distemper-, egg-tempera-, oil colour, imprimatura (in oil), oil-resin-colour, beeswax, linseed oil varnish, oil-resin varnish and a white ground on rabbit glue) in different combinations are investigated.

The category of combined experiments is represented by tests to determine the creep behaviour on the three-wood species. Specimens are cut from each wood species measuring 40 mm long and 15 mm × 15 mm as cross-sectional area. The specimens are loaded by constant mechanical compression stress by dead weight in length direction for two weeks and unloaded for two further weeks. The deformation in loading direction is regularly observed photogrammetrically and evaluated by digital image correlation. The experiment is conducted for the specified wood types, for the radial, tangential and longitudinal material direction, for constant relative humidity (RH) at 50% RH, 65% RH and 85% RH and changing RH between 50 and 85%. The experimental setup contains ten test stations, where one specimen is mechanically unloaded, three specimens are loaded at about 20%, three specimens at about 40% and three specimens at about 60% of the compression strength.

The validation experiments are conducted to investigate the more complex coupled behaviour of wooden specimens at hygro-mechanical loads. For the paintings of the replicas, the characteristic hygro-mechanical material-behaviour of wood and the interaction of wood and coatings is tested in the so-called panel strip experiment, where the deflection as a result of different relative humidities is measured. In this experiment, the different coatings and paint layers are applied to wooden stripes (spruce and lime) measuring 50 mm long × 250 mm wide × 4 mm thick. The wood qualities of the individual panels are selected to be almost constant and worked out from the same raw material. Slight variations in annual rings and mark positions have to be accepted. For the experimental run in the climate chamber, the axis of the pith and the axis of the supporting points on the test stand are marked on the strips. A spring gauge is positioned in the centre of the strip at the pith axis, which detected the deformations during the climate changes. In the climate scenario with constant temperature (20 °C), the strips are first conditioned to 85%, then to 35% relative humidity and finally dried to zero moisture content. For each setup, three stripes are measured.

Two further experiments are conducted: one experiment for moisture induced swelling pressure behaviour[3] and one for moisture induced shrinkage tension

---

[2] According to DIN 52184.

[3] The experiment is conducted according to experiments in (Krauss 1988).

behaviour.[4] For the first experiment, specimens with 60 mm length and a cross-section of 15 mm × 15 mm are created of each wood species with radial or tangential material direction in length direction. The expansion of the samples is restricted in length direction. For the second experiment, specimens, measuring 60 mm long × 5 mm thick × 20 mm wide at the ends and 15 mm wide in the middle in length direction, are produced, also with radial or tangential material direction in length direction. Shrinkage is restricted for the length of 20 mm in length direction of the specimens. In these experiments, the wooden samples are exposed to an increasing or decreasing relative humidity while swelling and shrinking is prevented in one direction. The induced stresses are measured by a load cell.

Finally, the replicas are constructed according to the originals. For the four panel paintings, two supporting systems (two of each) on the back side are built for evaluation. A classical cradle system from the nineteenth century and a modern supporting system from 2015 on the base of the system by Ray Marchant and Simon Bobak[5] are modelled. The cupboard replicas are rebuild based on a very simple constructed type of cupboard, which is made in duplicate, but without the cornice attachment. For the panel painting as well as for the cupboard, the characteristic paintings and coatings are identified and applicated in a simplified way and with a slightly different stratigraphy compared to the original one. To validate the structural behaviour, the deformation of the panels and the cupboard is measured in two different climate conditioning scenarios. During a climate conditioning phase of four weeks, the replicas are loaded by very high and low relative humidity in a climate chamber (85% RH and 25% RH). In the second phase, a long-term conditioning, the replicas are stored on an uninsulated attic under approximate outdoor climatic conditions for the duration of a seasonal cycle. The deformation is measured by a photogrammetrical system at regular intervals, using coded and uncoded measuring marks. A software evaluates photos taken from the replicas and calculates coordinates and displacements of the marks.[6]

2.2 Development on FEM Models

For the numerical investigation of wooden cultural heritage objects, several material models are used to describe the material behaviour (Reichel 2015). The mechanical behaviour is described by elasticity and plasticity models for the short-term behaviour, visco-elasticity and visco-plasticity models for the creep behaviour and interface models for discrete cracks with stiffness and strength reduction during crack-opening and a contact formulation for pressure loading. The other part of the numerical modelling represents the moisture transport. Due to the cellular structure of wood, two phases of moisture are described: bound water and water vapour.

---

[4] The experiment is conducted according to experiments in (Koponen and Virta 2004).

[5] For more details of the restoration of the cradle see (Van Grevenstein et al. 2014), p. 51.

[6] The "Aicon 3D Studio Software" by the current manufacturer HEXAGON is used for this purpose.

Both phases are transported through wood by diffusion processes. Beside diffusion, sorption is modelled. At the surfaces, water vapour is taken from or given to the surrounding air. A resistance to water vapour emission is modelled for a boundary layer and coatings by surface elements.

For the numerical simulation of the replicas, geometrical models are created and the components identified influencing the structural behaviour. Then, the models are discretised by the finite element method and the previous material formulations are associated with the finite elements. For the panel painting with the old cradle system, solid elements are created for the panel and the cradle system, taking into account the different wood species and material directions. For the coating, surface elements are used with mechanical behaviour tangential to the surface and water vapour transport through the surface, considering the permeability of the paint layer. Surface elements without mechanical properties are attached at the other surfaces for the moisture interaction with the surrounding air. Between the movable parts of the cradle and the other wooden parts, interface elements with a contact formulation perpendicular to the interface are attached.

3 Results

Only some of the results of the experiments, simulations and validations can be presented here as an outlook on a future comprehensive publication.

3.1 Diffusion Test

One exemplarily result of the diffusion tests with the permeability of the coatings is depicted in Fig. 1. Differences can be seen for different coatings, but also for different relative humidities. An exponential function is proposed to model the relative-humidity-dependent permeability of the coatings, using the parameters a and b for the dry permeability and moisture dependency. The differences become clear in the bar charts. The layers with the highest water vapour flow resistance are beeswax and the resin-containing coatings. The largest moisture dependency shows beeswax, too.

3.2 Stripe Experiment

During the stripe experiments, the influences of the different layers of paint and coatings on the wooden materials become visible, especially the difference between the two types of wood, spruce and lime. It appears that the interaction behaviour is similar in each individual experimental set-up, but the intensities of the reactions

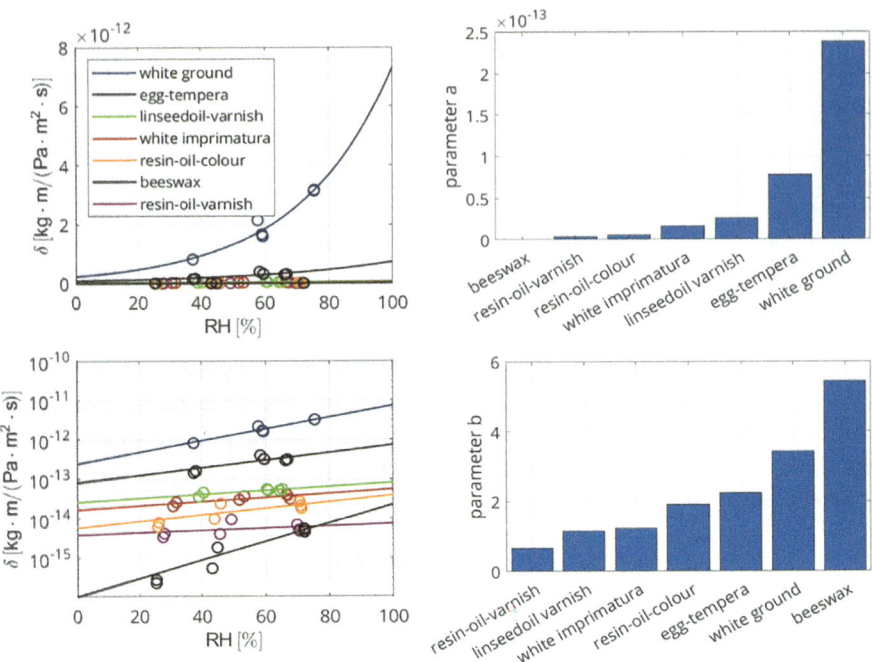

Fig. 1 Diffusion properties of the different binder media

shows scattering. The deflection can be split into different parts caused by different reasons. The limewood stripe without varnish shows deflection (convex on the core side) due to material inhomogeneity caused by the annual rings of the wood and the difference of stiffness and swelling- and shrinking behaviour between wood and coating. The limewood stripe with tight varnish shows additionally deflection due to a moisture gradient, which occurs, because the moisture uptake is fast at one surface and very slow at the coated surface. Since the moisture gradient vanishes with time by diffusion, also this part of deflection vanishes. The spruce stripe (concave on the core side), having lower strength than limewood, shows also plastic deformation.

3.3   Replicas

The measurements on the replicas have not yet been completed at the time of the conference. Only the first climate phase can be exemplarily evaluated here using the example of a panel with the historical cradle. The diagram on the top left-hand in Fig. 2 shows the relative humidity driven in the climate chamber and corresponding wood moisture content of the panel at different measuring points. The diagrams in Fig. 2 show the deflection of the panel painting in length direction in a cross-section

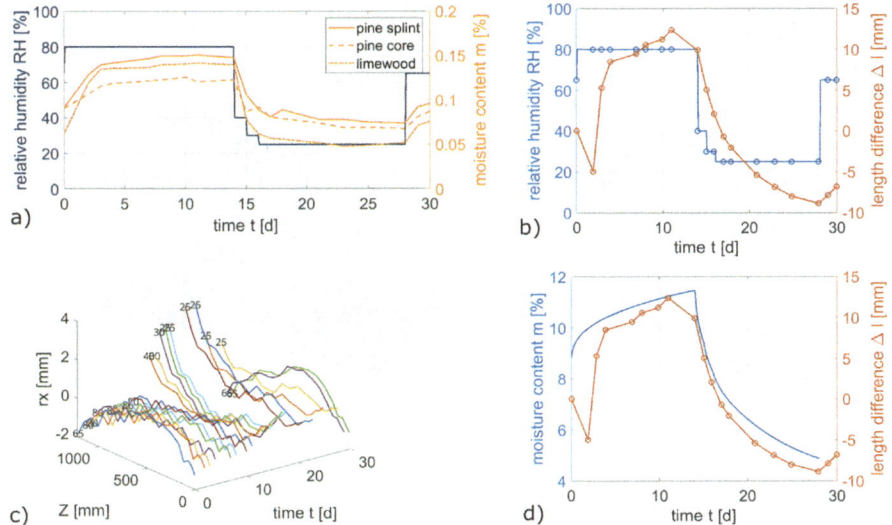

Fig. 2 Different results of the panel measurement, **a** relative humidity in the climate chamber and moisture content in different parts of the panel painting, **b** regulated relative humidity and measured length difference versus time, **c** deflection of the panel due to moisture change, **d** simulated moisture content and measured length difference versus time

of the panel, which is caused by material inhomogeneity and moisture gradient, as described before. A visible effect can be identified also from the cradle system at the back side in the jagged graph of each measurement epoch. The diagrams at right-hand (Fig. 2) show the length difference of the panel due to swelling and shrinking depending on the moisture content. The diagram to the top right (Fig. 2) shows the time-dependent relative humidity in the climate chamber in correlation with the measured length difference and below right-hand (Fig. 2) related to the simulated mean moisture content in the panel. A clear correlation is visible between the simulated moisture content and the measured length difference.

The simualtions of the panel paintings, which predict the climate scenarios of the climate chamber, show the following summarised results (Fig. 3). The visible effect of the movement of the sliders inwards and outwards the horizontal cradle part in the humid and dry climate respectively is remarkable. It can be observed both in the simulation and for the real replicas. On further examination of the simulations, the stresses caused by drying can be seen and visualised, especially in the coating. This result allows an estimation of the risk of damage in the coating layer. It is also possible to show the concentration of bound water. The influence of moisture on the wood panel itself as well as on the paint layers can thus be evaluated and visualised. It becomes clear that the complex structure of the panel with the massive old cradle leads to thicker and thinner cross-sections of the panel and, thus, to different moisture storage capacity and different high moisture content areas within the panel. This in turn creates tension stresses within the wood structure.

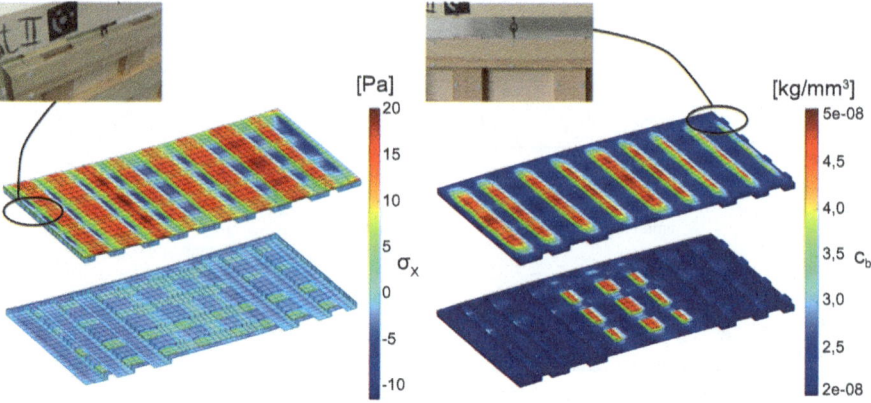

Fig. 3 Results of the simulation (front and rear perspective): left, stress in the panel's length direction of x in wet and, right, visualisation of the bound water concentration in dry climate

4   Conclusion and Outlook

At the current stage of the research work, it can be stated that the material parameters for wood and coatings necessary for the simulation development have been investigated in order to provide the numerical models with reliable data. The diffusion properties of various characteristic historical painting layers are investigated. The results will be published separately in cooperation with the Virtex research project.[7]

Furthermore, climate induced stresses in wood and the interaction of wood and coatings have been studied, as well as the influence of different coatings on the behaviour of coated wood. The results of the preliminary tests and small sample experiments developed in the project will also be published in the near future.

Within the results of the preliminary experiments, the structural behaviour of a panel painting is studied and simulated. The scenarios presented under climatic stress are supplemented by long-term experiments, which have not yet been completed. Exciting measuring and validation results are expected from the comparison of the two supporting structures (historical and modern) as well as from the comparison of the functioning and an artificially non-functioning prepared historical parquets.

Through the finite element models developed, it is possible to predict the structural behaviour of wooden artwork exposed to different boundary conditions realistically. The mechanical conditions can be evaluated and, thus, the risk of damage to the wood structure and paint layers can be estimated. For an adequate risk assessment, specific damage criteria must be applied to coatings. Furthermore, with the current state of research, material uncertainties and inhomogeneities still need to be taken into account.

---

[7] Project 100371102, funded by the State Ministry of Science and Cultural Affairs of Saxony: Virtual experiments for artwork—Virtex.

In the interdisciplinary discussion on the application of numerical simulations, it is often indicated by conservators, that more detailed and comprehensive material descriptions and parameters need to be considered for predictions that are more accurate and a practice-oriented application for conservation and restoration science aspects. Essentially, the material properties of all materials used in the production of historical cultural objects, historical as well as contemporary conservation and restoration materials and their interaction need to be investigated, including also ageing and degradation effects on the material properties. In addition, it has to be examined, how object-specific history and restoration measures can be considered in the simulations. The panel painting by Cranach the Elder for example is supposed to exhibit plastic deformations and damage due to the historic cradle system, which has to be taken into account in a simulation of the original panel. The state-determination of historic objects, the requirements for the numerical tool, the investigation of material parameters of original objects as input for the numerical tool are only possible in an interdisciplinary cooperation between conservators and engineers. For this purpose, an intensive interdisciplinary communication with conservation science is planned.

SPONSORED BY THE

Federal Ministry
of Education
and Research

Acknowledgements The authors gratefully acknowledge the support of the work by the Federal Ministry of Education and Research (BMBF) within project 03VP05781 "Numerical analysis tool for the simulation and risk estimation of climatic and mechanical loaded works of art made of wood—CULTWOOD".

References

Koponen, Simo, and Jari Virta. 2004. Stress relaxation and failure behaviour under swelling and shrinkage loads in transverse directions. In *COST E35 workshop I, vila real, Portugal*, 1–4.

Krauss, A. 1988. Untersuchungen über den quellungsdruck des holzes in faserrichtung. *Holzforschung und Holzverwertung* 40: 65–72.

Reichel, Susanne. 2015. *Modellierung und simulation hygro-mechanisch beanspruchter strukturen aus holz im kurz- und langzeitbereich*. Dresden: Dissertationsschrift, Institut für Statik und Dynamik der Tragwerke, TU Dresden.

Van Grevenstein, Anne, Britta New, Christina Young, Kate Seymour, Roger Groves, and Velson Horie. 2014. *The conservation of panel paintings and related objects: Research agenda 2014–2020*. The Hague: Netherlands Organisation for Scientific Research.

Refinements on Accelerated Corrosion Testing—Development of Sustainable, Reproducible Indicator Plates

Alexandra Jeberien, Nivin Alktash, Sabrina Maric, and Bernd Szyszka

Abstract The Material-Checker project (MAT-CH) covers the most common corrosion pollution test, the so-called Oddy test, which has been widely performed to examine material used in museums and collections. Besides many advantages, the test is also known for its limitations regarding inconsistent equipment, varying procedures and subjective evaluation leading to ongoing revisions. Project MAT-CH aims to innovate the Oddy test and thus hopes to contribute to a higher standard in pollution control. While the project initially focused on the reaction vessel, recent focal point emphasises on the development of sustainable indicator plates. In opposition to the bulk metals of the Oddy test, MAT-CH indicators will consist of thin metal-coated glass substrates, which lead to enhanced precision, replicability, and sustainability. The paper describes the project objectives, as well as the development process and evaluation of samples specifying layer thickness, density, homogeneity, and surface roughness. Furthermore, the examination of prototypes, including performance and in-situ tests, as well as results from microscopy and instrumental analysis are discussed.

1 Accelerated Corrosion Testing and Objectives of Project MAT-CH

The impact of pollutants on museum collections and archives has been well-known in heritage conservation. Next to air pollution, construction material and display cases contribute to indoor air quality. Hence, materials should be examined prior to use.

The Material-Checker project (MAT-CH) has been exploring the most common corrosion test. The so-called *Oddy test* has provided an easy to perform analysis method for museums staff worldwide in almost fifty years. The test bases on the

A. Jeberien (✉) · N. Alktash · S. Maric
HTW Berlin, University of Applied Sciences, Berlin, Germany
e-mail: alexandra.jeberien@htw-berlin.de

B. Szyszka
Technische Universität Berlin, Berlin, Germany

© The Author(s) 2025
Á. F. Perles-Ivars et al. (eds.), *Collection Care*, Springer Proceedings in Archaeology and Heritage, https://doi.org/10.1007/978-3-031-85655-6_13

corrosion caused by volatile organic compounds to silver, copper, and lead under accelerated conditions, like elevated humidity and temperatures (Oddy 1973). To generate reactions, plates of all three metals, a material sample and a mini-test tube filled with distilled water are inserted into a laboratory glass, which is closed with a silicon stopper. The whole set-up is then placed in a hot cabinet at 60° Celsius for 28 days (Green and Thickett 1995). After reaction time all results are visually examined and compared to a control sample. Depending on the level of corrosion the examined material might be categorized for permanent or temporary use (no or slight oxidation) or proofs to be unsuitable (severe corrosion) for museums purposes (Thickett and Lee 1996).

The *Oddy test* delivers comprehensive and, provided accurately executed, reliable results for a whole range of compounds. It has been successfully applied to examine materials used in storage, transport, and display. Though, it is also known for its scientific limitations, including inconsistencies in equipment, procedures, and especially evaluation, all leading to restricted replicability. For this reason, the test has been subject to ongoing revisions (Bamberger et al. 1999; Tsukada et al. 2012; Wang et al. 2011). MAT-CH project also focuses on described drawbacks, but specifically aims to innovate the equipment, thus hoping to contribute to a higher reproducibility. The project is a collaboration between the Technical University Berlin and HTW Berlin.

After improvements on the reaction vessel during the first part of the project (Heine and Jeberien 2018), MAT-CH 2.0 focuses on the development of reproducible and sustainable indicator plates. According to the original and latest protocols (Korenberg et al. 2018; Green and Thickett 1995; Oddy 1973), plates need to be cut from metal foil and manually cleaned from oxidation residues with glass fibre pens. While this procedure causes inaccuracy, cross-contamination and due to abrasion particles severe health threads, it especially leads to imprecise results. In opposition, MAT-CH indicators consist of corrosion-stable borosilicate glass, which is machine-cut and metal-coated using plasma technologies. This automated production results in ready-to-use indicators, thus leading to more precise and comparable test results. At the same time thin-film coating enhances sustainability, given that the use of scarce resources is significantly reduced.

2 Development Process of Reproducible Indicator Plates

The reproducible MAT-CH indicators were developed in cooperation with the Institute of High-Frequency and Semiconductor System Technologies (HFT) of the Technical University Berlin.

As a basis for the indicators, corrosion-stable borosilicate Corning® EAGLE XG® glass was chosen, cut to size, and drilled for hanging indicators in the reaction vessel.[1] For coating, high-quality copper and silver tube targets, and one lead foil target, each with a purity of >99.99% were selected.[2] Finally, shadow masks with a coating area representing the surface of the indicators (20 × 20 mm), as well as holders to fix the substrate and shadow mask during the coating process were size manufactured.

Silver and copper indicators were to be produced using gas flow sputtering technology (GFS). Compared to conventional sputtering technology, GFS operates at much higher process pressures in the mbar range (1 mbar = 100 Pa). To avoid contamination of the equipment, a separate device was chosen for lead indicators (high-vacuum magnetron sputtering system). For the coating process, a glow discharge is ignited on the target, resulting in surface erosion. Subsequently, the target metal is transported towards the glass surface via forced convection in an argon flow and is deposited as thin metal films (Gudmundsson 2020; Paduraru et al. 2003).

The development process started with a series of 100 nm layers for silver and copper on normal laboratory glass. Hereby, basic coating parameters such as power, dynamics, and pulse mode were tested. Evaluating the results, those samples produced with high power and in pulse mode already showed reasonable layer thicknesses, while samples manufactured in pulse mode and by dynamic coating (DC), resulted in better homogeneity and adhesion. However, silver layers were less homogeneous, therefore the process power was reduced on silver. In parallel, 100 nm lead layers were coated on Corning glass, varying in pulse mode, dynamic coating, and in pressures from 0.3 to 1.2 Pa. Again, those samples coated in DC mode adhered well, while high pressure coating resulted in better resistivity.

The results were implemented in the production of 100 and 200 nm lead indicators. At the same time, 100 and 200 nm silver and copper indicators with the given parameters were also produced, followed by an investigation of the general corrosion behaviour of all samples. The corrosion test revealed the need for further optimization, especially on the lead coatings. Though, the other indicators were also aligned. For lead, the process pressure was revised once more, resulting in a high-pressure and DC mode manufacturing procedure. In opposition, silver and copper indicators were still produced in pulse mode. Eventually, a selection of 200 nm medium-power coated silver, 100 nm high-power coated copper, and 200 nm high-pressure coated lead indicators was chosen for the performance test (Figs. 1 and 2).

---

[1] Glass substrates are 1,1 mm thick. Silver and copper are coated together on one squared substrate, sized 25 × 25 mm. On the other hand, lead is fitted on a separate glass, sized 12.5 × 25 mm. All substrates have a drilled hole with 5 mm in diameter in the top part of the plate.

[2] The tube targets (Ag, Cu) are each 60 mm long, with an outer dimension of 50 mm and inner diameter of 40 mm. The foil target (Pb) has a diameter of 75 mm and a material thickness of 6 mm.

3   Performance Test of MAT-CH Indicators

After a first selection of MAT-CH indicators had been developed, their general corrosion behaviour and overall functionality were examined. In this regard, two performance test series were designed. Both series included prototypes of MAT-CH indicators and the known bulk metal plates (Green and Thickett 1995). Together with
reaction triggering sample material MAT-CH indicators were positioned in the novel
reaction vessels (Heine and Jeberien 2018), respectively the metal plates in standard laboratory glass tubes (Thickett and Lee 1996). All samples were brought to
accelerated reaction, using the hot cabinet, elevated humidity, and high temperatures.

3.1   Pre-test on Corrosion Behaviour

The corrosion test included almost similar copper and silver indicators, which were
produced with high to medium power and in pulse mode. In opposition, two variants
of lead prototypes were examined. Both variants were produced with high power,
but either in DC or in pulse mode. All variants came in 100 and 200 nm thickness.

The corrosion test was carried out under accelerated conditions, aiming for 100%
humidity and temperatures of 60° Celsius, in the innovated MAT-CH reaction vessels.
Three samples for each variant were examined, plus one control sample each (total
= 16). In order to draw reliable conclusions regarding the corrosion behaviour, it
was opted for selected test material leading to explicit reactions on the metal layers.
Therefore, white *KAPA Line*, a rigid PUR foam board, which releases acetic acid
and leads to strong reactions on copper and lead, and black sponge rubber (EPDM
foam), with a corrosive effect on silver due to sulphur-crosslinked production, were
chosen to accelerate reactions.

In parallel, the corrosion rate was analysed in an extra set-up, using the known
equipment of laboratory glass tubes and silicone stoppers. For the documentation of resistance and transmission the experiment was supplemented by an in-situ
measuring device.[3] Indicators according to the above parameters were chosen (total
= 4), as well as the same test material to trigger corrosion. Both test settings, the
MAT-CH reaction vessels and the in-situ measurement were placed in the hot cabinet
for four weeks.

After reaction time, all indicators were visually evaluated revealing highly diverse
results: while the copper and lead coatings showed various corrosion products,
ranging from dark, iridescent colouration and powdery deposits to completely
exhausted metal layers, the reactions of the silver indicators were much more subtle.
In comparison, the in-situ measurements correspond to these results. The strongest
corrosion activity was documented on the copper sample, which presented a gradual

---

[3] The multimeter Fluke 8808 was connected with cables, which were passed through the silicone
stopper of the laboratory tube and attached to the metallic surfaces of indicators with conductive
silver lacquer.

increase in resistance. In contrast, hardly any resistance was recorded on the silver sample until the third week.

3.2 Test on Functionality

Whereas the first performance test concentrated on the general corrosion behaviour, the second test series focused on the overall functionality, such as comparability, validity, and replicability of the new indicator plates.

The functionality test was performed in a comparative test setting of MAT-CH and Oddy equipment and was repeated three times. While MAT-CH indicators were prepared in the novel reaction vessels, bulk metal plates were fit in conventional glass tubes, enabling a parallel reaction. Five samples for each indicator metal, plus in each case one control sample, were examined per run. This gives a total of 3×15 results per setting and nine control samples.

The functionality test involved three reaction triggering test materials: a vulcanized rubber labelled as *gikoZell ZGA* (unsuitable), the PVC rigid foam board *Forex classic* (temporary) and *Tyvek®* soft fleece (permanent). These materials were primarily chosen according to the criteria of multiple testing, overall reaction caused by the material, and the corrosion potential for each metal. This way the three evaluation stages for each metal layer could be reached systematically. The selection was based on the project team's own corrosion test data file, and compared to other data available from the British Museum and the AIC Conservation Wiki.

After completed reaction time, results were visually evaluated on a macro- and microscopic level. The first impression revealed homogenous corrosion patterns amongst all samples of each metal. Furthermore, corrosion patterns representing all levels from permanent to unsuitable, emerged from the test runs. Finally, all test runs gave comparable results between MAT-CH and Oddy set-up, meaning that similar corrosion patterns were visible in both settings.

All in all, the functionality test delivered constant results for all three test runs. The sample material classified as unsuitable (rubber material *gikoZell ZGA*) caused moderate to strong reactions on the metal coatings, including dark bottom areas on copper, as well as pitting on silver and copper. The temporary classified PVC foam board *Forex classic* also resulted in the darkening of the lower parts, though on silver, while copper layers were almost completely covered with a yellow-greenish discolouration. The lead indicators turned opaque only. In connection with the suitable classified test material *Tyvek®* soft fleece, copper layers showed slight darkening too, and lead indicators demonstrated slight transparency again. Silver layers remained unchanged with this material.

4 Analytical Evaluation of MAT-CH Indicators After Performance Tests

For the investigation of the reactivity behaviour and examination of corrosion products, instrumental analysis such as UV–VIS, ellipsometry, and SEM–EDS was performed on selected MAT-CH indicators (Figs. 3 and 4). While the spectroscopy determined the interaction between transparency, reflectivity and absorption, the optical methods of ellipsometry and scanning electron microscopy helped to further characterize material and layer properties of the indicators. Finally, corrosion products were examined using energy-dispersive X-ray spectroscopy.

To receive comparative data for the UV–VIS, freshly coated silver and copper indicators were examined and resulted in high reflectivity and low transparency. On the contrary, the oxidized permanent, as well as the corroded temporary samples stayed rather reflective, while their transparency only slowly accumulated. The reason for this behaviour, hence the missing decrement of reflectivity especially on temporary silver samples might be due to slow reaction with the material *Forex classic*. Only unsuitable samples showed a distinct reversed ratio, in that reflectivity declined, and transparency clearly increased.

Subsequently, the characteristics of the coatings were explored with ellipsometry. The resulting measures do not offer much information by themselves, thus need to be related to model calculations. For the characterization of MAT-CH indicators a two-phase model was applied, showing layer resistivity ρ (Ωcm) and roughness. As an example, calculations on the post-tested silver indicators resulted in thick and attached-to-the-glass layers with low resistance. Above that, thin layers with high p-values were detected. The surface of the latter might be described as a mixture of silver and air and represents the roughness. The layer also increases from permanent to unsuitable samples, thus indicators tested with the unsuitable material *gikoZell ZGA* had the highest roughness.

Additionally, scanning electron microscopy was conducted before and after the performance tests. Compared to freshly coated copper indicators, measurements of permanent samples only displayed little differences, while significant alterations were visible on both temporary and unsuitable indicators. Particularly on cross-sections, these observations are in line with the two-phase model from the ellipsometry, meaning that the upper layer increases in thickness (roughness) with corrosion level, respectively on temporary and unsuitable samples. The same applies for silver indicators, showing major differences between freshly coated metals and temporary to unsuitable samples, while strong alterations are already visible on the permanent samples of lead indicators. However, this might be due to higher reactivity of the lead coatings in general.

To further determine visible corrosion, elemental analysis was performed using energy dispersive spectroscopy (EDS). Measurements on the control sample and permanent copper layer indicated low carbon and some oxygen content. In comparison, the temporary sample singled out reduced oxygen, while carbon slightly increased. Additionally, nitrogen appeared, which was even more apparent on the unsuitable sample. On the unsuitable layers notable values of sulphur were detected, also the oxygen had sharply risen. Both, the occurrence of nitrogen and sulphur are most likely related to the triggering sample material. While nitrogen serves as a blowing agent in the manufacturing process of PVC foam boards (*Forex classic*), the production of rubber material (*gikoZell ZGA*) often includes sulphur. The latter most likely causes dark colouration on copper, as detected on the bottom rim of unsuitable samples. The EDS analysis of the unsuitable silver samples revealed similar results, in that an increased oxygen content was present, and nitrogen appeared as well. Finally, the measures on control and permanent lead samples showed strong oxygen contents, which underlines the theory that no corrosion, but oxidation had occurred, although strong alterations are visible on all samples.

In summary, instrumental analysis was able to confirm visual observations received from the performance tests of MAT-CH indicators. Though, the examination highlighted the necessity for refinements, especially on lead indicators.

5 Status Quo

The Material Checker project (MAT-CH) innovates the equipment of the well-known accelerated corrosion test—the *Oddy test*, for pollution control in museums. The project aims to enhance usability, replicability, and sustainability, thus hopes to contribute to a more reliable method.

The paper described the need for innovation, the development process and examination of reproducible, ready-to-use indicator plates. Although, the development process was planned and operated thoroughly, performance tests on the novel indicators have once again made it clear that further research is inevitable and would lead to a more explicit outcome.

Future optimizations will have to include broader coating series of lead indicators to explore and understand process parameters, especially the impact of pressure on layer formation. Another examination of silver and copper layers regarding thickness also seems to be reasonable. Since these optimizations need to be tested for corrosion behaviour and functionality, performance tests might be performed with different choices of material samples.

Finally, another MAT-CH project shall include objective evaluation methods for the classification of corroded indicators (Figs. 1, 2, 3, and 4).

Fig. 1 Innovated reproducible MAT-CH indicator plates: Ag (l.), Cu (m.), Pb (r.) layer

Fig. 2 MAT-CH indicator plates after the functionality test using polyethylene soft fleece: Cu (l.), Ag (m.), Pb (r.) layer

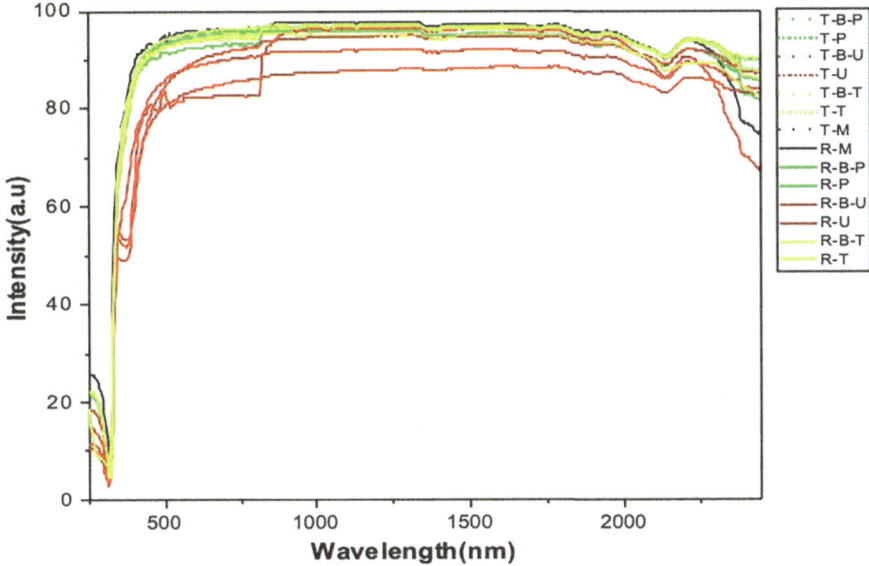

Fig. 3 UV/VIS spectra of silver indicators: freshly coated (M), permanent (green), temporary (yellow), unsuitable (red) rating

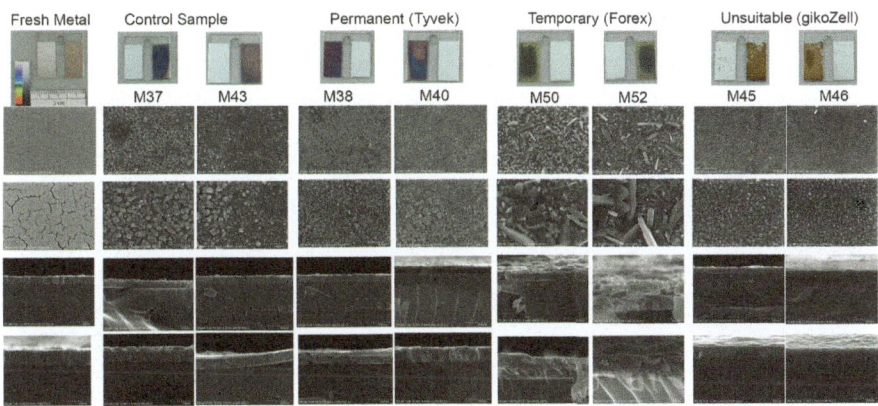

Fig. 4 Comparison of SEM analysis, showing surface and cross-section images of freshly coated, as well as control, permanent, temporary and unsuiteable copper samples

References

Bamberger, Joseph A., Howe, Ellen G., and Wheeler, George. 1999. A variant Oddy test procedure for evaluating materials used in storage and display cases. *Studies in Conservation* 44: 86–90.

Green, Lorna, and Thickett, David. 1995. Testing materials for use in the storage and display of antiquities—a revised methodology. *Studies in Conservation* 40: 145–152.

Gudmundsson, Jon Tomas. 2020. Physics and technology of magnetron sputtering discharges. *Plasma Sources Science and Technology* 29: 113001.

Heine, Hildegard, and Jeberien, Alexandra. 2018. Oddy test reloaded: Standardized test equipment and evaluation methods for accelerated corrosion testing. *Studies in Conservation* 63: 362–365.

Korenberg, Capucine, Keable, Melanie, Phippard, Juie, and Doyle, Adrian. 2018. Refinements introduced in the Oddy test methodology. *Studies in Conservation* 63: 2–12.

Oddy, William Andrew. 1973. An unsuspected danger in display. *Museums Journal* 73: 27–28.

Paduraru, C., Belkind, Abraham, Becker, Kurt, Lopez, Jose L., Delahoy, Adam, and Guo, Sheyu. 2003. Pulsed DC, gas-flow hollow cathode discharge: A source for sputter-deposition. In *46th SVC Annual Technical Conference Proceedings*, 130–134. San Francisco, CA: Society of Vacuum Coaters.

Thickett, David, and Lee, L. R. 1996. Selection of materials for storage and display of museums objects. *British Museum Occasional Papers* 111: 24–26.

Tsukada, Masahiko, Rizzo, Adriana, and Granzotto, Clara. 2012. A new strategy for assessing off-gassing from museum materials: Air sampling in Oddy test vessels. *American Institute for Conservation of Historic and Artistic Works* 37: 1–7.

Wang, Sheng, Kong, Lingdong, An, Zhisheng, Chen, Jianmin, Wu, Laimin, and Zhou, Xinguang. 2011. An improved Oddy test using metal films. *Studies in Conservation* 56: 138–153.

Design of Tables with Unified Criteria for the Preliminary Phases of the Study of Cultural Heritage: Risk Indicators and Priorities

Tanja Mastroiacovo, Maria Pilar Soriano Sancho, and José Luis Regidor Ros

Abstract This investigation presents the results derived from the design and use of registry tables drawn up according to unified criteria on the preliminary phases of diagnosing cultural heritage, presenting the advantages, limitations and improvement margins obtained after the first application experiences. This project emerged from the need to find a useful operating method to be applied in conditions with limited resources and a large number of works or elements. The aim is to provide an objective approach to comprehension of the general and specific conservation state of said works, thus allowing users to prioritise and organise the diagnosis evaluations and specific analyses to be performed in upcoming phases, to facilitate the channelling of specific economic and technical resources, based on the potential derived from studies from the statistical science field and the contributions of the official glossaries generated for the field of conservation and restoration of cultural heritage. The strategy presented here is formulated for the creation of conservation risk indicators. The cases described, consisting of stone materials and mural paintings, constitute the first step towards the development and enhancement of the use of registry tables with unified criteria, offered as a basis for future implementation and improvement.

1 Introduction

This research presents the experiences matured following the development and implementation of tables with unified criteria, specifically dealing with those derived from their application to heritage ensembles composed of stone material and mural paintings. The main axis of this study was developed around the problems found through

T. Mastroiacovo (✉)
Research and Cultural Heritage, Tanja Mastroiacovo Art Studio & Conservation, Brescia, Italy
e-mail: research@tanjamastroiacovo.com

M. P. Soriano Sancho · J. L. Regidor Ros
Universitat Politècnica de València, Instituto Universitario de Restauración del Patrimonio, Valencia, Spain

© The Author(s) 2025
Á. F. Perles-Ivars et al. (eds.), *Collection Care*, Springer Proceedings in Archaeology and Heritage, https://doi.org/10.1007/978-3-031-85655-6_14

different professional experiences, in which the approach and assessment phases of the conservation state of various cultural assets were limited in operational terms due to the plurality of elements to be examined and the minimal availability of professional, technical and time resources.

The relevant aspects considered when structuring the methodological proposal were focused on the initial phases of evaluation and diagnosis of cultural heritage. In these stages, the main actions carried out by the conservation professional can be classified as follows: observe, register, organise, estimate/quantify, analyse and interpret. The effectiveness of the observation and appreciation tasks of the alteration and deterioration phenomenology present in cultural assets is intimately connected to the registry methodology, constituting the base of a correct understanding of its conservational needs. This has highlighted the relevance of the selection process for the information to be registered in the preliminary inspection phase, determining which phenomena to direct the professionals' attention to. The data organising media and the registry worksheet model useful for registration and analysis of the alterations and decay present in the study subjects, from their visible effect, was inspired by statistical techniques, in particular those employed in biodiversity studies.

The method developed was conceived to be employed with a meagre resource investment, and the predefined objective of registering and organising the information, thus contributing to the prioritisation of specific activities to be implemented in the latter phases of diagnostic studies and evaluations. Establishing a methodology for the data gathering represented the core of the operating section, due to the abundance of variables, factors and materials that influence the conservation of cultural heritage. This highlighted the need to design a table for recording information with unified criteria, whereby the approach to and understanding of the specific conservation and study requirements of the groups of works could satisfy the need to generate potentially comparable data on different scales.

The use of standardised resources, compiled from the official glossaries of the cultural heritage conservation sector, has favoured the structuring of the concepts being registered in the data matrix. The contributions of the statistical studies have been adapted to the demands of the conservation sector, offering a new strategy for the registration and analysis of information. The registry tables with unified criteria were formulated to record and provide quantitative and qualitative data, organising them thanks to the generation of numerical values defined as *conservational risk indicators*. The tool, designed to tackle the initial study phases, aims to offer guidelines to prioritise subsequent phases, facilitating the orientation of the deeper actions and/or specific diagnostic analysis to come.

2 Objectives

The aim of the method presented here consisted in the design of a data gathering and analysis strategy, by means of creating tables with unified criteria, useful for the management and resolution of problems inherent to the preliminary phases of study and diagnosis of cultural heritage.

To accomplish this goal, these specific objectives were presented:

- Design a matrix model with unified terminology for registering the general and specific conditions of study groups, considering a minimal resource investment.
- Implement the designed method in real cases and problems to determine its viability, advantages and limitations.
- Analyse the qualitative and quantitative information provided by the method and its usefulness, according to the means used in the registration and analysis operations of the given information.

3 Methodology

The operational methodology has been structured in different phases (Fig. 1) and consists of methods' design, implementation of said method, and result evaluation.

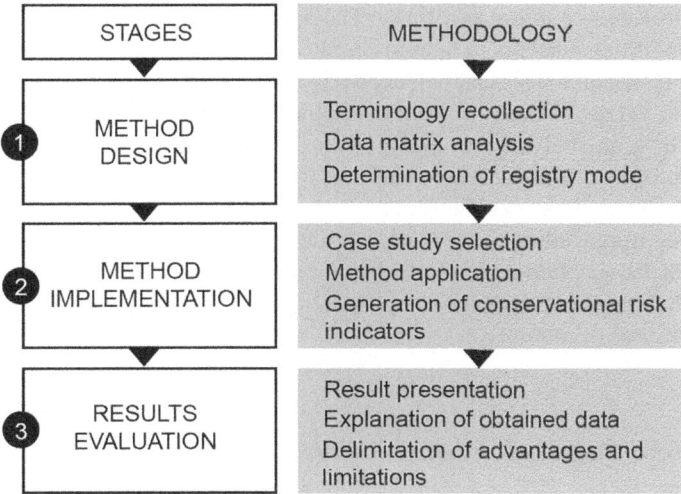

Fig. 1 Methodology followed in this investigation

3.1   First Stage—Method Design

3.1.1   Objective

Design a matrix model for the specific cultural heritage information registry that allows a clear, fast and representative data collection.

3.1.2   Methodology

a. Compilation of standardised terminology from the glossaries of the stone material and mural painting conservation sector to provide the method with recognised and internationally employed technical terminology.
b. Analysis of the data matrix employed in the statistical techniques to structure the method for the collection and information study.
c. Identification of the derived potential of compliance of a registry table, accompanied by a quantitative estimation of the affected surface by alterations and/or deterioration.

The compilation of specific terms for alterations and deterioration made reference to the official glossaries ICOMOS-ISCS (2008), Illustrated Glossary on Stone Deterioration Patterns, for stone material, and HORNEMANN INSTITUT (2015), European Illustrated Glossary for Conservation Terms of Wall Painting and Architectonic Surfaces, for mural paintings. The selection of these texts as a basis for organisation of the concepts to be registered complied with the aim of relying on a unified and internationally acknowledged terminology and the possibility to consult the semantic values of each of the terms, along with their graphic representation. Parallel to this, an analysis was made of the structure of the data matrix of information employed in the Ecology field and in exploratory data analysis (Instituto de Investigación de Recursos Biológicos Alexander Von Humboldt 2014), with the purpose of comprehending the applicative potential derived from statistical science, in pursuit of its adaptation to the demands and resolution of the problems which characterise the preliminary phases of diagnosing cultural heritage in conditions of minimal resource availability.

3.2   Second Stage—Implementation of the Method

3.2.1   Objective

Apply the registry table designed for the data recollection of real case studies in order to evaluate advantages, limitations and improvements on the use of the proposed method.

3.2.2 Methodology

a. Selection of case studies belonging to the class of mural painting and stone material works.
b. Application of the method to the case studies selected by means of the registered phenomenology of the alteration and deterioration in the designed data matrix.
c. Generation of conservation risk indicators and analysis of the quantitative and qualitative information acquired.

For the application of the method to study groups composed of stone material, sculptures and decorative elements were chosen from various temple facades located in San Luis Potosí (S.L.P., Mexico), whereas for the implementation on mural paintings, the decorative part of the interior courtyard of the Peotillos Hacienda was chosen, located in Villa de Hidalgo (S.L.P., Mexico). Each study group was chosen for its accessibility and amplitude, in terms of the number of constitutive elements and/or surface extent.

Each element of the groups is paired with an alphanumerical coordinate as a reference (M1, M2, M3, etc....) and its specifications are gathered on an identification sheet, where its useful characteristics are broken down for ease of location and identification.

The data matrix was organised to estimate, in the registry phase, the superficial extent (%) of the alteration and deterioration phenomenology visible in the cultural asset, using numerical values ranging from 1 to 4, referencing the data matrixes used in the Ecology field. The notation of values in this study was done on a photographic basis.

The data and numerical values thus obtained were elaborated through a mathematical calculation using Excel© software, in order to generate global (study group) and specific (study sample) conservational risk indicators. The collection structure, compliance mode of the worksheets and elaboration of the information proposed were organised to deliver both qualitative information (alteration and deterioration present) and quantitative (impact and extent), in order to identify areas or elements with further conservative instability along with the spread of its problems. Finally, the resulting information was interpreted referencing the glossaries used for the table design, looking to bring a cause-effect correlation that will allow evaluation of the information obtained and its representativeness according to the real conservational context.

3.3 Third Stage—Result Evaluation

3.3.1 Objective

Identify advantages, limitations and viability of the proposed method.

3.3.2 Methodology

a. Presentation of the results obtained during the structuring and application of the designed method.
b. Explanation of the information gathered through the method and its usefulness according to the registry operations and analysis of the alteration and deterioration phenomenology present in the case studied analysed.
c. Delimitation of the advantages and limitations found according to the viability, practicality, effectiveness and representativeness criteria of the information obtained.

In the third stage, the experiences and results obtained through the research are compiled; in order to compare advantages, limitations, operating potential and possible margin for improvement. The assessment of the method and the discussion of the results was generated thanks to the presentation and comparison of information, such as: data obtained, resources employed, quality of the information, interpretation of results, usefulness and depth level achieved through the development of the information and the means employed.

4 Results

From the specific alteration and deterioration phenomenology data that affect cultural heritage, worksheet prototypes were generated in which the criteria were organised into two axes (Table 1) where the vertical axis unites the types of alteration and deterioration, while the horizontal axis breaks down the case studies through alphanumerical coordinates, associated with an identification sheet.

The terminology compiled for each of the artwork groups was the result of the contributions made to the cultural heritage conservation sector by institutions involved in the generation of glossaries with unified terms. These documents brought an optimal starting point for the structuring of the data to be gathered in the registry tables (Tables 2, 3), providing the system with an array of terminology known and accepted in the international scope. The publications employed for this purpose

Table 1 Registry worksheet structure prototype

Types	Samples		
Alteration or deterioration	M1	M2	M3
Classification A	1–4	1–4	1–4
Classification B	1–4	1–4	1–4
Classification C	1–4	1–4	1–4
Etc.	1–4	1–4	1–4

were those from the HORNEMANN INSTITUT (2015), European Illustrated Glossary for Conservation Terms of Wall Painting and Architectonic Surfaces, y las de ICOMOS-ISCS (2008), Illustrated Glossary on Stone Deterioration Patterns.

The compliance mode of the data matrix was formed in terms of abundance, employing numerical references from 1 to 4, in which the values corresponded to an evaluation of the extent of the phenomena (Fig. 2). These parameter references were established with the aim of providing a medium to institute an approximate evaluation of the range of the alteration and deterioration, in order to delimitate its extent in the registry phase.

In the implementation phase of the method, the following activities took place:

- Photographic documentation;
- Alteration/deterioration phenomena documentation;
- Data analysis;
- Generation of conservation risk indicators;
- Graphic representation of the information.

According to the cultural assets studied, the means used for the method application (Table 4) can be considered minimal, with scant deployment of economic resources.

The advantages and limitations found during the experience in application of the method were analysed according to the criteria proposed in the research approach:

- Viability of use according to the necessary resources for its application (temporal, human, economic and technical).
- Practicality of the method according to the accessibility and clarity of the information required for its use.
- Effectiveness and representability of the information obtained, evaluating the quality of the results in relation to the means necessary for its implementation.

After analysing the information, thanks to the data generated, it was possible to:

- Identify the alteration and deterioration phenomenology present along its qualitative and percentage estimation (Table 5);
- Generate a conservation risk indicator (Table 6) of the cultural heritage groups, both in a global (collection) and specific (piece or study area) character;
- Graphically represent the information (Fig. 3);

Table 2 Detail of the registry sheet formulated for Stone materials

Type	Group	Characteristic	M1	M2	M3
Detachment	Disintegration	Crumbling	1–4	1–4	1–4
		Granular disintigration	1–4	1–4	1–4
		Powdering, chalking	1–4	1–4	1–4
		Sanding	1–4	1–4	1–4
		Sugaring	1–4	1–4	1–4

Table 3 Detail of the registry sheet formulated for mural paintings, Peotillos Hacienda

Specifications	M1	M2	M3	M4	M5	M6	M7	M8	M9	M10	M11	M12	M13	M14	TOT
Features induced by material loss—**Lacuna**	2	2	1	2	2	1	1	1	1	1	1	1	1	1	18
Features induced by material loss—**Cavity**	1										1				2
Features induced by material loss—**Crumbling**	2	2	1	2	2	1	2	2	2	1	1	1	1	1	21
Features induced by material loss—**Abrasion**	2	2	1	1	1	1	1	1	1	1	1	1	1	1	16
Cracks and deformation—**Cracks**	1	1	1	1			1								5
Cracks and deformation—**Static cracks**		1	1												2
Cracks and deformation—**Deformation**	2	2	2	2	2	1	1	1	1	1	1	1	1	1	19
Cracks and deformation - **Craqueleure**	2	2	2	2	2	1	1	1	1	1	1	1	1	1	19
Chromatic alterations and deposits—**White veil**			1	1											2
Chromatic alterations and deposits—**Yellowing**						1	1	2		1	1				6
Chromatic alterations and deposits—**Darkening**		2				1	1								4
Chromatic alterations and deposits—**Whitening**							1								1
Chromatic alterations and deposits—**Bleaching**	1	1	1	3	2	1	1	1	1	1	1	1	1	1	17
Chromatic alterations and deposits—**Salt efflorescence**			1	1											2
Chromatic alterations and deposits—**Overpaint**			4		1									3	8
Chromatic alterations and deposits—**Deposits**	1	1													2
Anthropic interventions—**Structural Interventions**	1	1	4	1	1		1					1	1	1	12
Anthropic interventions—**Historic interventions**			4										4	3	11
TOTAL	15	17	24	16	13	8	12	9	7	7	8	7	11	13	18

METHOD DESIGN

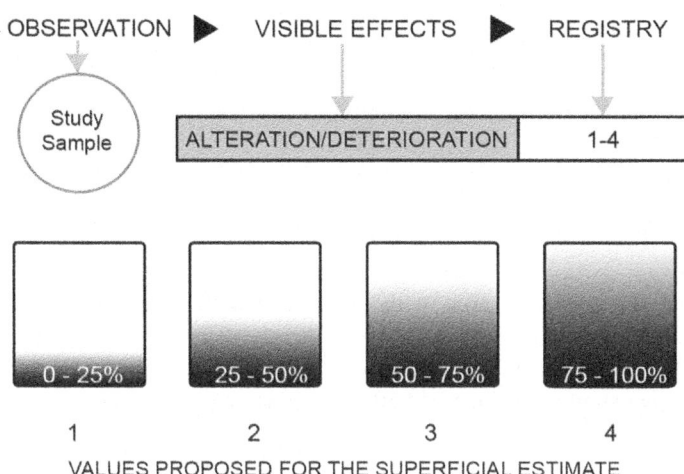

Fig. 2 Parametric reference proposed for the sheet

Table 4 Cultural assets analyzed and resources

Case studies	Resources			
	Temporal		Technical	Professional
	Data registry	Data Analysis		
Stone scupltures (34 elements)	2 h 25 min	2 h 25 min	Registry tables Photographic camera Computer with Excel©	1 Conservator Restorer
Mural paintings (14 areas)	2 h	1 h		

Table 5 Information generated after the entry and data analysis, Metropolitan Cathedral

Metropolitan cathedral—sample M10 (amount of damage: 9)

Alteration and damage		Data record estimation	Damaged area (%)
Typology	Specification		
Discolouration and deposit	Dirt/soil	4	75–100
Discolouration and deposit	Black crust	3	50–75
Discolouration and deposit	Concretion	2	25–50
Discolouration and deposit	Crust	2	25–50
Discolouration and deposit	Deposit	2	25–50
Disintegration	Sugaring	1	0–25
Features induced by material loss	Loss of matrix	1	0–25
Features induced by material loss	Roughening	1	0–25
Features induced by material loss	Material loss	1	0–25

Table 6 Deterioration indicators obtained after the analysis of the gathered information, San Miguel Arcángel Temple

Sample	Conservational risk indicators		
	Plurality (occurrence n°)	Extent (Surface %)	Global (n° + %)
1	16	21	37
2	11	12	23
3	10	11	21
4	1	1	2

- Determine the piece or area with a larger risk indicator and prioritize its conservation (Fig. 4);
- Identify the alteration and deterioration present in each work, piece or area;
- Estimate the extent or surface (%) affected by each phenomenon;

The information gathered was analysed using an Excel© sheet, with the necessary mathematical formulas designed to obtain quick numerical values, generating specific conservational risk indicators of (Fig. 5):

- Plurality of alterations and deterioration registered.
- Superficial extent of the observed alteration and deterioration phenomena.

The sum of all these has been defined as global conservational risk indicator and priorities, because it organises and gathers the information on the occurrence and extent of the collections phenomena.

Substituting the inconvenience determined by the lack of coordination between the official monitoring methodologies and the registry of heritage conservation conditions that can be adopted in different local realities, Eppich and Garcia Grinda (2015), the method presented in this study could favour the construction of efficient indicators. These, basing themselves on a unified terminology, will allow for a future centralisation and standardisation of information, thus providing the advantage of being able to compare and collate data on a different scale.

From the classification provided by the Getty Conservation Institute (2007b), in Guiding Principles—Recording, Documentation and Information Management for Conservation of Heritage Places: Illustrated Examples, the necessary resources for the application of the proposed method belong in level A, at low costs. This tool needs a minimal investment of time, human and technical resources. A few hours on site with the personnel in charge of the registry equipped with basic tools for the photographic documentation and sheets for the registry phases, and a computer with Excel© for the information analysis will be enough time. However, the quality of the information provided allows users to obtain B level data, of medium accuracy, contributing to the determination of the initial conditions of the cultural asset alongside a description of the existing problem, a general photographic registry and qualitative and quantitative reference information about the alteration and deterioration phenomenology present. The extents of the method, thanks to the possibility of being

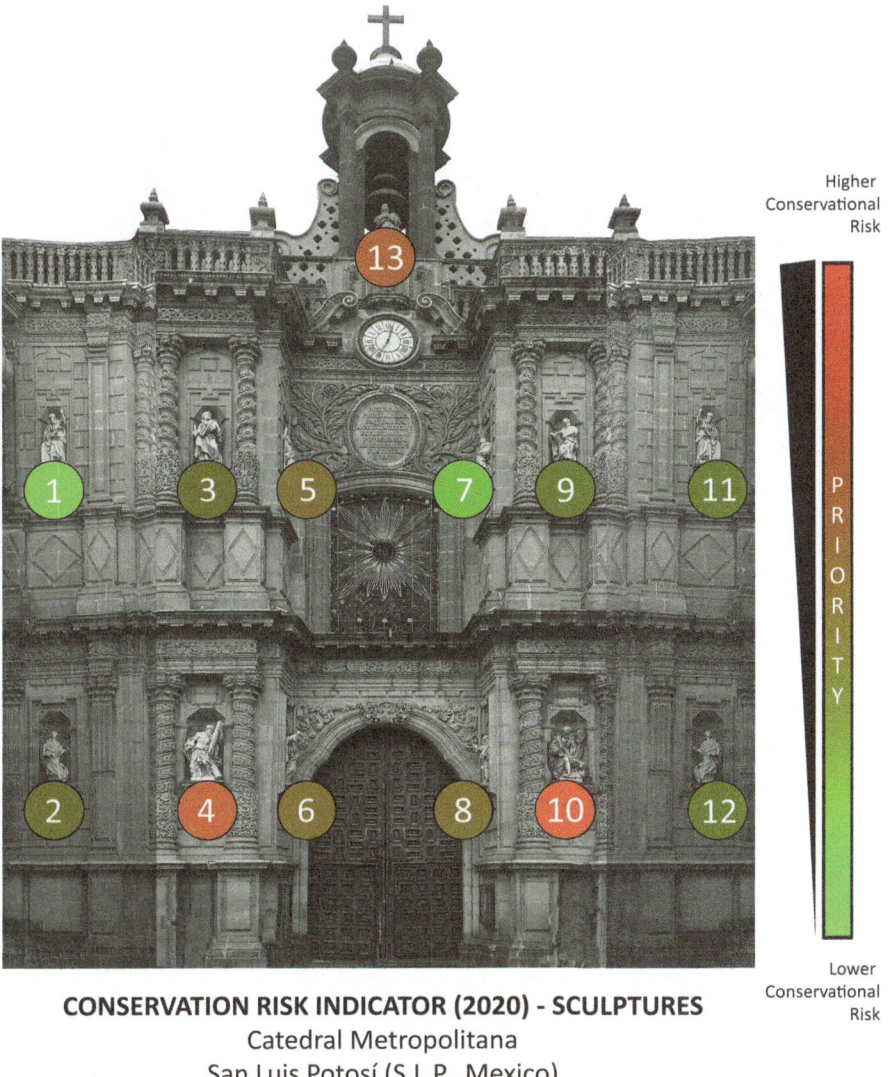

CONSERVATION RISK INDICATOR (2020) - SCULPTURES
Catedral Metropolitana
San Luis Potosí (S.L.P., Mexico)

Fig. 3 Graphic representation of conservation risk indicators, Catedral Metropolitana

easily incorporated into the maintenance and monitoring routines of heritage groups submitted to conservation and restoration interventions cover up to level C, thus meeting various operating needs.

These data, under the expert eye of a conservator, offer the opportunity of making a general sweep of the state of a cultural asset, while allowing a specific approach to each work/study sample, in favour of organising and directing analytical resources and/or specific studies.

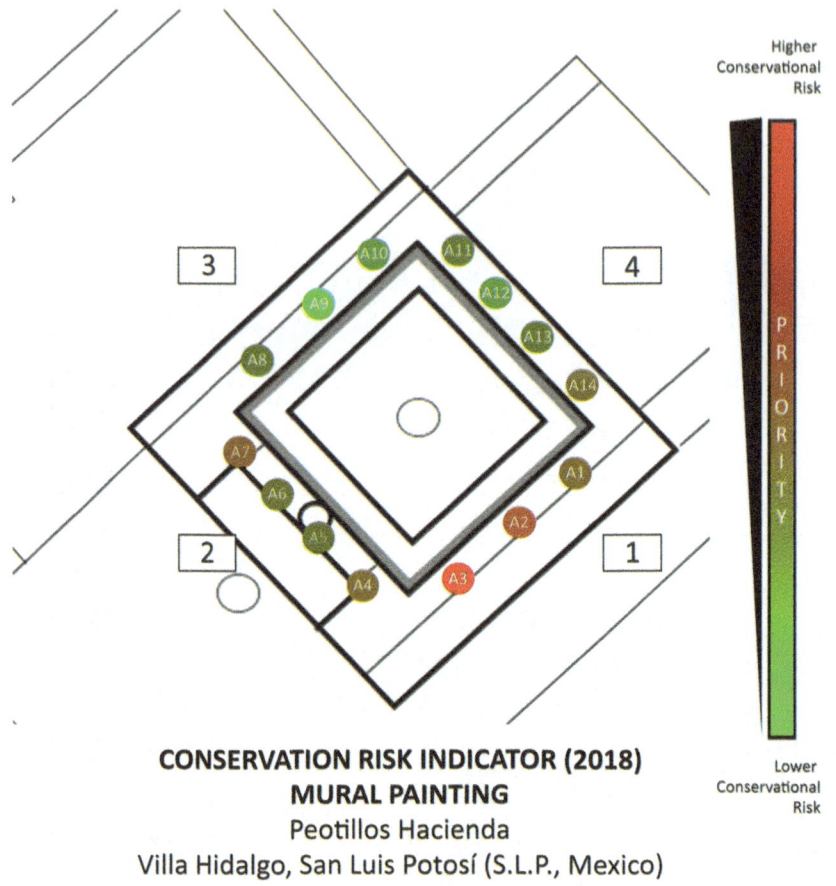

CONSERVATION RISK INDICATOR (2018)
MURAL PAINTING
Peotillos Hacienda
Villa Hidalgo, San Luis Potosí (S.L.P., Mexico)

Fig. 4 Graphic representation of conservation risk indicators, decorative murals in the Peotillos Hacienda

5 Conclusions

Considering the cases analysed in San Luis Potosí historical centre (Mexico) and the Peotillos Hacienda (Villa Hidalgo, S.L.P., Mexico) and the quantity and quality of the information generated in relation to the means employed for implementation of the method, it can be affirmed that the use of unified tables in the preliminary diagnostic phases of the cultural heritage has proven to be a useful and viable tool, capable of bringing affordable means of exploration to the conservational state of the collections. According to the meagre means necessary for its application, the proposed tool can easily be employed in circumstances or contexts characterised by economic, temporal, technical and human constraints.

Fig. 5 Conservational risk indicators generated after the data registry and analysis

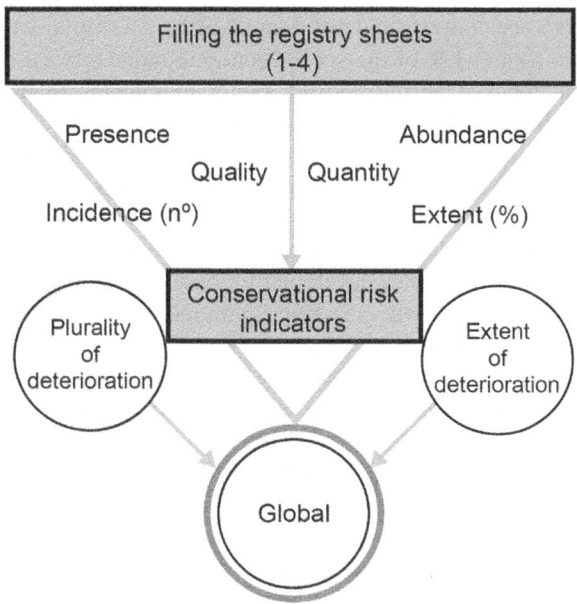

The fundamental and most complex phase of the presented study consisted of structuring the information and the design of the registry tables, because it is from them that the quality of the data given by the method derives. The design of the data matrix with unified criteria, inspired by contingency frequency model tables and structured around the two axes, has allowed the finding of representative results, extracting hypotheses and useful conclusions for the comprehension of the study contexts.

Regarding the demands for standardisation and individualisation of cultural heritage documentation, it is important to remember that the preliminary study phases depend on the organisation and effectiveness of the conservation actions, Mora et al. (1999) and Getty Conservation Institute (2007a). These are defined in relation to the complementary and contradictory aspects, which fluctuate between the need to thoroughly study each case, opening up to the comprehension of their own conservative, technical and historical peculiarities, and the will to drive the use of criteria that favours the comparison between the different realities investigated. Therefore, the results obtained from this investigation have to be considered in terms of an improvement in the phases of approach, comprehension and management of the conservational demands of cultural heritage, and accordingly to the basic needs for its implementation.

Thanks to the design of this unified criteria registry, the possibility emerges of implementing in the future a statistical type analysis of the information. The basic parameters used for the data gathering phases, would allow the creation of a centralised system for the monitoring of the conditions of cultural heritage, which, adopted on different scales, will favour the creation of statistics and comparatives between contexts and different realities in a similar way as the ones generated within the framework of the socio-economic indicators set out by UNESCO-FSC (2009), but focused on the registry and/or monitoring of the conservation state of works within a territory.

There is an open way towards a review and perfecting of the structure of the terms proposed in the unified tables designed for this research, focused towards the creation of an improved system in terms of precision and representability of the data, which could also be extended to other classes of cultural heritage. Among these cited are the tables developed for easel painting collections and sculptures made in polychrome wood, currently submitted to a review and timely improvement in their own structures.

In light of the results obtained by this research, it can be affirmed that the design and implementation of unified registry tables for the resolution of the preliminary study and diagnosis phases of cultural heritage can represent a versatile and very valuable tool, which, thanks to its ease of use and the minimal resources needed, can examine in depth the state of conservation of a broad spectrum of cultural assets, thus contributing to the prioritisation of specific conservational actions. The method, furthermore, can be applied to exploration techniques and complementary documentation, such as remote diagnosis through drones, currently strongly developed, or in cases where the documentation cannot be drafted in person. As an example, let us cite the value acquired by digital media and working remotely, due to the sanitary contingency of Covid-19. Comparable to this, it may occur that, due to emergency circumstances or natural catastrophes, the professional in the field may see him/herself limited without the possibility of commuting to the place of study. The opportunity of working on photographic files and remote video represents a possibility of great value.

In conclusion, the possible incorporation of the proposed system to the monitoring and diagnosing routines, thanks to the affordability of the tools needed for its use, represents a real and viable proposal to be implemented on the majority of realities, participating in the study of conservational demands of cultural heritage in a simple, quick and affordable way.

References

Eppich, Rand, and Garcia Grinda, José Luis. 2015. Management documentation indicators & good practice at cultural heritage places. In *The international archives of the photogrammetry, remote sensing and spatial information sciences, volume XL-5/W7, 2015 25th international CIPA symposium*, 133–140. Taipei, Taiwan.

Getty Conservation Institute. 2007a. Guiding principles - recording, documentation and information management for conservation of heritage places. https://www.getty.edu/conservation/publicati ons_resources/pdf_publications/recordim.html. Accessed 15 May 2020.

Getty Conservation Institute. 2007b. Recording, documentation, and information management for the conservation of heritage places: illustrated examples. https://www.getty.edu/conservation/ publications_resources/pdf_publications/recordim_vol2.html. Accessed 15 May 2020.

Hornemann Institut. 2015. European illustrated glossary for conservation terms of wall painting and architectonic surfaces. https://www.hornemanninstitut.de/doi/2016ewa2.pdf. Accessed 19 May 2017.

ICOMOS-ISCS. 2008. Illustrated glossary on stone deterioration patterns. https://www.icomos.org/publications/monuments_and_sites/15/pdf/Monuments_and_Sites_15_ISCS_Glossary_Stone.pdf. Accessed 3 March 2016.

Instituto de Investigación de Recursos Biológicos Alexander Von Humboldt. 2014. Métodos para el análisis de datos: una aplicación para resultados provenientes de caracterizaciones de biodiversidad. http://www.bionica.info/biblioteca/HumboldtAnalisisDatos.pdf. Accessed 20 March 2019.

Martínez, Fernández, and Víctor Manuel. 2015. *Arqueo estadística, métodos cuantitativos en arqueología*. Madrid: Alianza.

Mora, Paolo, Mora, Laura, and Philippot, Paul 1999. *La conservazione delle pitture murali*. Milano: Compositori.

UNESCO. 2009. Framework for cultural statistics. http://uis.unesco.org/sites/default/files/documents/unesco-framework-forcultural-statistics-2009-en_0.pdf. Accessed 15 March 2019.

Storage, Moving and Re-storage of Municipal Oporto Collections

Maria Aguiar, Inês Ferreira, Mariana Teixeira, Carolina Barata, and Ana Cabral

Abstract Within a wider collections management programme conducted by the *Divisão Municipal de Museus* (DMM) from Oporto City Hall, a collaboration has been started with the Portuguese Catholic University (UCP) for reorganisation of the municipal storage facilities and to set up the process of transfering collections to permanent housing in a refurbished building. The main goals of the project are to prepare the collections to be transfered, organise two temporary storages, establish procedures for moving the collections between temporary and permanent spaces, organise the final storage space and support the setup of guidelines for preventive conservation. In parallel, several training actions in the field of preventive conservation have been organised, to provide updated knowledge to the entire team of the DMM and to standardise the procedures to be implemented. Interconnections created during the project enabled us to capitalise personal skills and empower the DMM team with competences to lead autonomous work. The three-year project started in January of 2020, only 2 months before the pandemic outbreak in Portugal, and required permanent adjustments to adapt to this unexpected context. The work was done in close articulation with the conservation team of the Museum of the City and also with the register and inventory staff.

M. Aguiar (✉)
School of Arts, Portuguese Catholic University, Porto, Portugal
e-mail: mcaguiar@ucp.pt

M. Aguiar · C. Barata
Research Center for Science and Technology of Arts—CITAR, Porto, Portugal

I. Ferreira · M. Teixeira · C. Barata · A. Cabral
Museu da Cidade, Câmara Municipal do Porto, Porto, Portugal

© The Author(s) 2025
Á. F. Perles-Ivars et al. (eds.), *Collection Care*, Springer Proceedings in Archaeology and Heritage, https://doi.org/10.1007/978-3-031-85655-6_15

1 Introduction

Until 2016, Oporto City Hall had a central storage space in a building belonging to a palace that was then sold. Since then, assets were transferred twice to provisional warehouses while Oporto City Hall searched for alternatives to keep its collections on a long-term basis with adequate conditions for conservation, access and study.

The choice for the permanent storage facility fell upon a pre-existing building that had been constructed to provide social and educational support for young people in the city, but which was abandoned a decade ago. The building, called—*Abrigo dos Pequeninos*—went through an architectural project that devised seven technical storage spaces for ceramics, textiles, paintings, metals, furniture, paper and diverse objects. It also considered a conservation laboratory, a quarantine room, a photographic studio, a research/study room, with documentation and staff rooms.

The construction works began in September 2021 and the conclusion of the renovation was estimated for early 2023.

2 Characterization of Collections and Storage Spaces

Municipal collections in storage come from the *Museum of the City* (https://museud acidadeporto.pt/) that comprehends, among other spaces, the *Casa Marta Ortigão Sampaio, the Extensão do Romantismo*, the *Extensão do Douro* and *Atelier António Carneiro*. Items in storage also include collections related to the poet Eugénio de Andrade, to the painter and collector, Vitorino Ribeiro, to the Measurement Gauging Center and to the *Casa de São Roque*. Other collections come through donations, in small or large scale and do not have a specific attributed location.

Typologies are quite diverse, ranging from textiles, costume accessories, furniture, easel paintings, books, drawings, prints, paper, photographs, clichés, photographic equipment, zinc etchings, basketry, ceramic, pottery, tiles, chandeliers, plaster sculptures, gilded and polychrome wooden sculptures, objects in alabaster, silver, glass, metal, ivory, and leather.

In 2018, such collections were moved to a temporary warehouse with no conditions for permanent housing and the great majority remained packed and not easily accessed. Adding to that, the documentation system was incomplete, in line to what ICCROM and UNESCO international survey in 2011 showed—1 in 4 museums had the same problem (ICCROM and UNESCO 2011).

Regarding other storage facilities apart from this central warehouse, the Museum of the City possess smaller spaces within (some) museums to keep their assets out of the exhibition spaces. However, these small rooms are already full and do not provide extra space for preparation of collections, for condition survey, etc. While waiting for the renovation of the *Abrigo dos Pequeninos* for permanent storage, the DMM had already foreseen the use of another space that could temporarily house some of the items before proceeding to the permanent space, thus allowing for a better management of both spaces and their collections.

3 Set the Plan While Dealing with Pandemic Constraints

Based on the RE-ORG method developed by ICCROM, UNESCO and CCI (ICCROM et al. 2018) an action plan was devised concerning the (re)-organization of both temporary storages. The other goals related to the preparation of collections for transport and transference to the *Abrigo dos Pequeninos* and the setup of preventive conservation guidelines required other approaches also due the timelapse of the project.

At the start of the project, staff at the main warehouse consisted of three people and just on the day before Oporto city lockdown, the two first conservator-restorers hired by the DMM began their functions. They did not have time to meet the colleagues that they would go on to work with.

After two months, the common goal that united the entire team around the project was affected by the worldwide pandemic that forced everyone to stop. The doubt lay not in the purpose that was outlined, but in the way we were all going to deal with the new situation that we had never experienced, as we did not know the contours, or how long it would last. The works that had been started had to be interrupted as they were all taking place in the storage spaces and access was prohibited to all staff.

In this first phase, all staff went home, museums were closed to the public and all those who worked in front office functions were exempt from these services, and had to be assigned to other functions. So, contrary to what is the most common scenario, there was a greater availability of human resources for the museum area, which is much more used to dealing with the opposite situation: the lack of people.

Priority was to choose the actions that could be carried out at home and to verify which resources were needed to carry them out. The need for computers was great, not only for the DMM, but for all municipal departments, as they were all going through the same difficulties, so it was necessary to carefully select their distribution. One of the uses in which they had priority was to update the inventory, with several people being responsible for updating the online database, Inarte® (https://sistem asfuturo.pt/index_en.html) and for checking aspects that would be important for the development of the project, such as listing which items had or did not have a photographic record. This type of information would serve to know how many items were left to be photographed and what resources and time were needed to do so.

Simultaneously with the selection of actions that could be carried out at home and considering that there was a greater number of people available within the DMM, a survey was carried out about personal skills that could serve the interests of the project. People with competences in design, languages or sewing were found, which proved to be very useful.

This was the moment to take advantage and review preventive conservation procedures, calling on all the spaces of the Museum of the City to collaborate. The greater availability of people to respond to these requests was possible because they were not busy with the daily commitments required by their functions in museums. It was possible to put together an important source of information that served as a basis, for example, for the new team of conservators-restorers that had just arrived, to plan training actions that responded directly to the needs felt by the people who dealt with the collections on a daily basis. On the other hand, the training actions were designed to meet the project's needs, providing people with the appropriate skills to develop the planned actions.

The various surveys that were carried out allowed us to verify which procedures were in use regarding preventive conservation activities and compare them with other practices, with the aim of standardising and defining those that would be the guidelines to adopt. For this work, it was possible to count on a larger work team that was also responsible for the literature review on each topic, the translation of publications (when necessary), as well as discussion and review of the final texts.

No one knew how long the pandemic would last. At the outset, everyone went home and only much later did people go to the sites, on a rotating basis and always with a reduced number of people, while the first phase of deconfinement was in force. At this stage, the work was programmed in such a way that people who had to deal with a certain collection would do so out of phase with each other or the various actions would be distributed among several people. Working documents (such as spreadsheets) were created with all the information considered relevant about the various collections, which were shared by everyone involved, with a person responsible for reviewing and updating them. A colour code was created that allowed the marking of situations related to the actions in progress, tasks to be carried out, completed, etc. and which facilitated the progress assessment for each collection. At this time, it was decided to hold hybrid meetings, in which some people participated by videoconference, while a small group was at the storage site. The system proved to be very useful to engage all the team in the work and to guarantee their safety.

At this stage, when it became possible to go to the storage spaces, albeit rotating, it was also possible to survey the storage structures present in the main warehouse and to measure, number and mark them. This preparatory work was very important for taking decisions about whether the existing furniture storage capacity was sufficient for the needs, whether it was adequate or whether more needed to be purchased. On the other hand, its marking allowed the location of the structures to be more easily found in the floor plans that were produced by the team member who revealed herself to have skills in design and digital tools.

With the reopening of public spaces, the teams were demobilised, as there was a need for frequent rotation of the team members, which required enormous flexibility in planning and setting goals.

4   Plan in Progress

Concerning the reorganisation of the main warehouse, its state was documented and the major threats affecting collections were identified (NPS 2005) as being related to building and space conditions (poor protection against light and weather extremes, occasional water infiltrations, no access or possibility to move objects, no circulation areas, large areas occupied by non-collection items and lack of storage space), collections (poor object documentation, no identification of furniture storage units, uncertainty of object location, pest evidence) and to furniture storage (the majority of collection laid down directly on the floor, insufficient and inadequate storage units and absence of padding or packing systems) (Fig. 1).

Regarding the second temporary space, it provided an additional 400 m^2 storage area and is located 6 km from the main warehouse. At the start of the project, the building was empty and needed to be prepared to serve the intended purpose. After assessing its conditions, improvements were made to block daylight sources, to avoid environmental fluctuations through the construction of a more impervious entrance area, to increase security systems by installing intruder and smoke detectors and requesting daily police surveillance, to control pests by an integrated pest management plan and to allocate permanent staff to carry out storage-related activities.

Fig. 1 View of part of the main warehouse before its reorganization

With this second storage in preparation, the action plan could finally take place and some of the threats posed to collections housed in the main warehouse could be addressed.

Priorities were set to deal with the most immediate threats, such as pest contamination and poor access to the collections. The physical reorganisation of the main warehouse could only take place after releasing areas to allow us to unpack and assess the collections. A layout for the main warehouse was defined, which included storage spaces, proper circulation corridors, work areas for the inventory, labelling, marking, photography, packaging/unpacking processes, quarantine and anoxia area, storage of equipment and materials (Elkin 2019) (Canadian Conservation Institute). Within this reorganisation, objects that did not belong to the Museum of the City were returned to their owners and display cases, while exhibition equipment and other non-collection items were also removed.

The art historian, inventory and conservation staff of Museum of the City selected the items that should be transferred to the new temporary storage site, based on the following criteria: quality, materials, fragility or condition state. This selection was essential to help calculate the areas needed to accommodate the works, to predict which were the most adequate storage structures, as well as to make a more rational management of the space. A principle underlying this selection was not to occupy the entire new temporary storage site with this first selection of items, as it was necessary to guarantee free space for the rest of the reorganisation process of the main warehouse.

5 Anoxia

The detection of pest infestation required treatment and the risk of contamination had to be contained, thus the DMM asked for support to the Divisão Municipal de Arquivo Histórico (DMAH) to use its anoxia chamber. However, the dimensions of the chamber only allowed small and medium objects to be treated as it was constructed for library and archive assets. Besides that, availability had to be shared among DMAH campaigns timetables. As it was planned to install an anoxia chamber at *Abrigo dos Pequeninos*, its acquisition was anticipated to allow the treatment of the contaminated objects. The acquisition of a detachable and portable anoxia structure was considered, instead of the rigid and compact structure initially foreseen, to allow its future dismantling and re-installing at the permanent storage. The chamber was designed to fit large objects, measuring 2,70 m height by 2,50 m width and depth. From March of 2020 to November of 2021, 297 objects were treated and in the last campaign it was possible to provide the service to other municipal departments that suffer from the same problem.

The conservation team working at the main warehouse received training to operate the chamber and to produce anoxia bubbles, in order to work autonomously. The methodology to identify, isolate, and monitor infestations was defined, as also the cleaning and documentation procedures. Based on the training received, the in-house conservation team produced an illustrated operating manual (DMM 2021a).

The decontamination process allowed the safe transference of furniture and other objects to the temporary storage, which between November of 2020 and October of 2021, received more than 3.700 items of furniture, textiles, costume accessories, tapestries, gilded and polychrome wooden objects, zinc engravings, chandeliers, and oil paintings on paper and copper.

6   Reorganization of Storages and Assessment of Collections

After releasing the main warehouse from part of its contents, it was possible to start its physical reorganization. Storage areas were devised taking into account space rationalization, the necessity of wide passages for circulation of people and objects and the avoidance of daylight entrances in the roof (IMC 2007) (Fig. 2). Floor plans occupied by furniture storage were drawn in both storage spaces.

During the process, it was necessary to calculate areas provided by the different typologies of furniture as to evaluate if there were enough storage surfaces for the collections and also, to decide how these typologies will fit in the space to accommodate all the items properly. This allowed to consider the possibility of re-adapting storage furniture and the acquisition of new units. All the decisions considered the

Fig. 2   View of part of the main warehouse after its reorganization

final transference to the permanent storage, so dimensions and characteristics of the new-purchased storage furniture were taking that in consideration (ICOM 2012).

Storage furniture has been numbered and different typologies received a colour code to facilitate future tracking of object's location on the floor plans that were designed for this purpose.

The physical reorganization of the main warehouse was done while unpacking collections, thus it was necessary to predict object dislocations and temporary settlements for them. The strategy was to start unpacking the more fragile collections, as ceramics, earthenware, and glass, as to reduce the number of dislocations and potential accidents during boxes movements.

Collection's assessment followed a sequence based on a preliminary documentation survey that crossed data from manual card records with the database Inarte® that included checking the register(s) number(s), the need to group objects together, the dimensions of each object, the condition state, the marking and labelling methods, the existence of photographic records, among other information. In general, it was found that there were many inaccuracies in the inventory, namely in the typologies and in associations of sets and composites/elements. As the items were packed, this reviewing process could only take place after the entire collection become visible and, for instance, (un)group elements of a set, depending on each situation.

The registration and inventory staff participated in the process and helped in the final decision-making process. A relevant contribution was provided by two external specialists on ceramics and pottery that helped to clarify issues related to provenance, date, function, quality, manufacturer, marks, technical and material characteristics, and elements of a set. A later technical visit to the pottery museum (Museu da Olaria), in Barcelos (Portugal) guided by the museum director reinforced the valorisation of such collection and contributed for current and future networking.

After unpacking and reviewing data of the 1813 items of ceramic and pottery, the other collections followed the same paths. However, depending on the condition state and size of each collection, different approaches were taken to make it feasible on time—small collections were more easily renumbered and updated, such as the 119 gilded and polychrome religious wooden objects, while the totality of the ceramics/pottery, the 3083 textiles (up to now), the 651 metal, 529 glass items and others collections will wait for the moving to the permanent storage to be marked with the new register number, for instance. Until then, they have received acid-free labels to help their current and future identification (Buck and Gilmore 2010).

The 81 metallic drawers with documents, photographs, engravings, postcards, and drawings will be reviewed later at the *Abrigo dos Pequeninos* because they include much more than the 1097 pieces registered in Fig. 3. The collection of textiles is quite large (3083 items are being reviewed) and there are still many costume accessories to be addressed. A similar situation occurs with furniture, where 184 pieces from a total of 345, are documented. The 430 zinc etchings require initiation of the documentation process as they do not yet possess, a register number.

Thus, decisions are being made according to the work demand of each collection and the awareness of the level required to be continued after moving. Until November of 2021, 97% of the items became visible and more accessible which allowed to carry

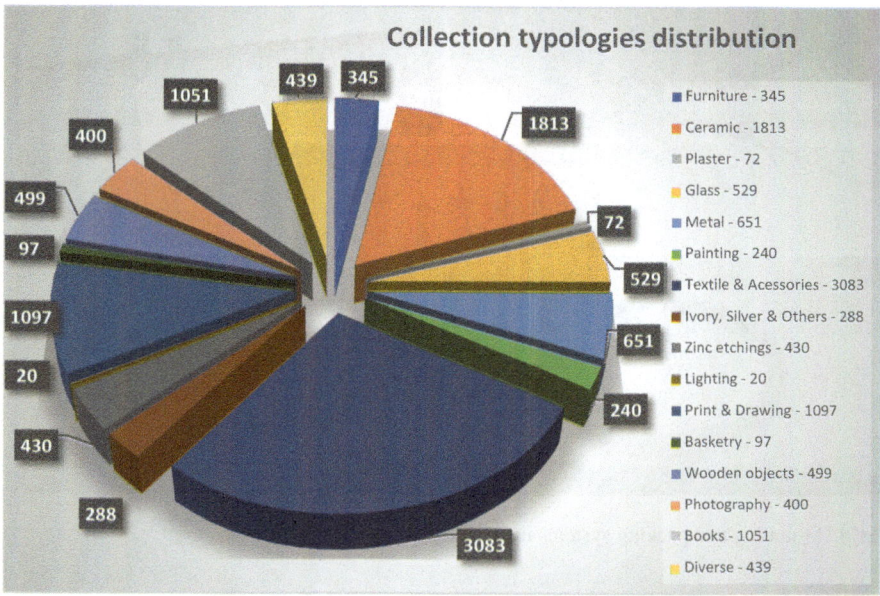

Fig. 3 Pie chart showing the collections distributed by type and the corresponding number of items

on photographic campaigns, condition state surveys and start a plan on mounting and packing systems for storage and for moving.

7 Packing and Moving

Regarding packing and mounting systems, the deadline and resources allocated for moving to the new refurbished building do not allow to treat all the collections at the same level. Only a small percentage of the collections had permanent or temporary boxing or mounts, thus the amount of work required is overwhelming. During the project, 68 cotton dust covers for chairs and 38 for textiles have been produced, which added to the existing ones, allowed the protection of 85 hanged costumes (Robinson and Pardoe 2000).

The goal is to protect the most vulnerable collections or objects and provide adequate systems for permanent storage while others would receive a temporary packing to be further developed after being moved.

Sustainability of the process is one of the aims to be achieved through the reduction of the materials used, the creation of standard packing models according to typologies, through systems that could facilitate examination and access of the objects and through the re-use of transport containers properly padded and size-shaped to accommodate the permanent boxes/supports and mounts, during moving (Fig. 4).

Fig. 4 Mounting and packing systems in progress

8   Preventive Conservation Training Program

Throughout this process, a parallel inner training program has been developed by the conservation and inventory team leaders in order to provide updated knowledge to the entire team of the DMM and to standardize the procedures to be implemented and the materials to be used concerning labelling and marking (NML 2010); environmental monitoring and control (Michalski 1993); integrated pest management (https://museumpests.net/) (Museum Pests.net.); cleaning; handling, packing and transporting (Illes 2006) (https://stashc.com/the-publication/).

For each topic, the surveys taken among museum´s staff helped to direct training to reality. Training sessions took place after the first lockdown and during the second one.

This first edition led to the writing of two manuals (DMM 2021b, c) made available to all services, which cross the specific needs of the Museum of the City and the most recent practices disseminated through bibliography and electronic resources provided by reference institutions in this field.

The practical sessions are still ongoing, some in mock-ups as the marking methods while others are being used to speed-up some actions needed in storage, as cleaning and packing.

The sessions regarding pest control and climate monitoring were developed in different spaces of the Museum of The City. This contributed to risk assessment processes and standardizing procedures regarding relative humidity (RH), temperature (T), light, integrated pest management and proposing continuous improvements.

After RH and T analysis, it was obvious that the main warehouse was not fitted for collection storage and that the second temporary storage can be improved with small adjustments to be categorized as class A (using ASHRAE specifications) (Michalsky 2007).

During the reorganization of the textiles collection, it became evident that some of the white textiles required treatment before they could be re-accommodated, used for exhibitions, for study or for other purposes. An external textile conservator-restorer was called for a conservative washing of a selection of pieces and to provide training to the conservation staff of the DMM that is taking care of the collection on the municipal storage.

9   Conclusion

Despite the enormous constraints due to the pandemic, there was a strong willingness to join forces so that the storage project could progress. The cessation of the museums' daily routine allowed time to reflect, discuss and define issues that would probably have taken longer to be addressed and would not have been addressed with in such a large team.

The survey of staff' skills, often extra-professional, played a major role, not only for the valorisation of this knowledge, but also for directing it towards the project goals, to which they ended up being associated after the deconfinement. The possibility to match personal interests and skills within a project is a key-factor to engage professionals and to progress much further due people´s involvement.

The interconnections created during the project were a chance to optimize personal skills and empower the DMM staff with competences through a purposed-designed training program, leading to a more collaborative, but also, autonomous work.

Up to now and within RE-ORG ten quality criteria, the ongoing project allowed to meet 5 of them: *one qualified member of staff is in charge; the storage rooms contain* (almost) *only collection objects; separate spaces are dedicated to support functions: office, workroom, storage of equipment and other materials (non-collection); no object is placed directly on the floor; objects are arranged by category.* A sixth could be considered—*every object is free from active deterioration*—as pest infestation was a major threat for the collections and it is controlled, but other types of damage must not be forgotten such as corrosion, for instance.

The protection offered by the storage and the building was also improved after the partial moving to new temporary storage and will improve much more, after the transference for the requalified permanent storage. The key policies and procedures need to be adapted to this new context and the inner training plan, followed by the writing and dissemination of the labelling/marking and preventive conservation manuals already provide a good basis.

The project is being a valuable chance to gain a much deeper knowledge about these collections behind scenes and to provide relevant data to support decision-making within a wider collection management program.

Acknowledgements To all staff of the Museum of the City and Divisão Municipal de Museus of the Câmara Municipal do Porto

To the Conservation and Restoration Center and to Research Center for Science and Technology of the Arts of the Portuguese Catholic University

References

Buck, Rebecca A., and Jean A. Gilmore. 2010. *MRM5: museum registration methods*. Washington, DC: American Association of Museums Press.

Canadian Conservation Institute. https://www.cci-icc.gc.ca/resources-ressources/. Accessed 12 October 2020.

DMM. 2021a. *Guia prático para a utilização da câmara portátil de anoxia: Tenda e bolhas, 2021*. Porto: Internal Procedures of Divisão Municipal de Museus, Câmara Municipal do Porto.

DMM. 2021b. *Manual de operações de conservação preventiva*. Porto: Internal Procedures of Divisão Municipal de Museus, Câmara Municipal do Porto.

DMM. 2021c. *Manual de operações de marcação de peças*. Porto: Internal Procedures of Divisão Municipal de Museus, Câmara Municipal do Porto.

Elkin, Lisa, and Christopher A. Norris. 2019. *Preventive conservation: collection storage*. New York, NY: Society for the Preservation of Natural History Collections.

ICCROM and UNESCO. 2011. International storage survey. https://www.iccrom.org/section/preventive-conservation/re-org. Accessed 6 January 2020.

ICCROM, UNESCO, and CCI. 2018. RE-ORG method. https://www.iccrom.org/themes/preventive-conservation/re-org/method. Accessed 6 January 2020.

ICOM. 2012. Gestão de acervos em reservas museológicas. Informação ICOM-PT. http://icom-portugal.org/multimedia/info%20II-15_dez11-fev12.pdf. Accessed 16 September 2021.

Illes, Véronique. 2006. *Guide de manipulation des collections*. Paris: Somogy.

IMC. 2007. Plano de conservação preventiva: Bases orientadoras, normas e procedimentos. Temas de Museologia. http://www.patrimoniocultural.gov.pt/static/data/ljf/ipmplanoconservacaopreventiva.pdf. Accessed 2 February 2021.

Michalski, Stefan. 1993. Relative humidity: a discussion of correct/incorrect values. *ICOM Committee for Conservation* 2: 624–629.

Michalski, Stefan. 2007. *The ideal climate, risk management, the ASHRAE chapter, proofed fluctuations, and towards a full risk analysis model: Experts roundtable on sustainable climate management strategies*. Los Angeles, CA: The Getty Conservation Institute.

Museum Pests.net. https://museumpests.net/. Accessed 12 October 2020.

NML (National Museums Liverpool). 2010. National Museums Liverpool Guidelines on marking and labelling methods and positions. Collections Trust. https://collectionstrust.org.uk/resource/guidelines-on-marking-and-labelling-methods/. Accessed 3 April 2020.

NPS (National Park Service). 2005. The museum handbook. Part I: Museum collections. NPS museum management program. https://www.nps.gov/museum/publications/MHI/MHI.pdf. Accessed 17 June 2020.

Robinson, Sarah, and Tuula Pardoe. 2000. *An illustrated guide to the care of costume and textile collections*. Scotland: Museums & Galleries Commission. Scottish Museums Council.

Storage Techniques for Art, Science and History. Investigating solutions. FAIC. https://stashc.com/the-publication/. Accessed 28 February 2020.

New Premises, New Challenges: Integrating Preventive Conservation at the National Library of Greece

Zoitsa Gkinni

Abstract The National Library of Greece was established in 1829, and in 1903 moved to *Vallianeio*, an imposing neoclassical building in the center of Athens. The building comprises of three wings, left and right wings serve as storage areas, whereas the center wing hosts the main reading room. Similarly to historical buildings, *Vallianeio* was designed to passively control the interior climate. For more than a century, reading rooms and working areas have been periodically heated or cooled to create a comfort zone for the occupants. In 2006, the Stavros Niarchos Foundation announced its plans to fund the development of the Stavros Niarchos Foundation Cultural Center (SNFCC), that included new facilities for the National Library, the National Opera and Stavros Niarchos Park. In 2016, SNFCC was the first large-scale cultural project to achieve LEED Platinum certification in Europe. The goal was to integrate active and passive technologies that make SNFCC one of the world's most environmentally sustainable buildings. In this context, the new facilities aim to provide suitable conditions for the collections' long-term preservation. In 2018, after the massive endeavor of collections' transfer from *Vallianeio* to SNFCC, the early preservation activities focused on achieving a smooth transition to the new climate and storage facilities. The goal set was to maximize the building's potential for efficient energy use. Starting with identifying and documenting the strengths and weaknesses of the storage areas (close and open bookstacks), the following step was to assess the indoor and outdoor environment, leading to taking measures and launching new projects for improving the collections' storage conditions. Within this new and challenging context, this paper presents the environmental documentation that was launched due to the transfer project and was enriched with further projects at the new premises, along with the reasoning and argumentation over collections' preservation planning demands and the actions taken so far to raise awareness and to classify preventive conservation as a critical fundamental action.

Z. Gkinni (✉)
Conservation Department, National Library of Greece, Kallithea, Greece
e-mail: zgkinni@nlg.gr

© The Author(s) 2025
Á. F. Perles-Ivars et al. (eds.), *Collection Care*, Springer Proceedings in Archaeology and Heritage, https://doi.org/10.1007/978-3-031-85655-6_16

1   Introduction

The National Library of Greece (NLG) was established in 1829. Since 1903 NLG was housed at Vallianeio, named after its donors. It stands as a landmark building in the city center, designed by architect Theophil von Hansen and supervised by Ernest Ziller. Vallianeio is part of the "architectural trilogy" at the center of Athens, along with the University and the Academia of Athens. For more than a hundred years, Vallianeio served its purpose and the public with its imposing reading room and closed bookstacks. However, collections' acquisitions and constant public demands led to the recruitment of more storage areas and services. In 2006, the Stavros Niarchos Foundation announced its plans to fund the development of the Stavros Niarchos Foundation Cultural Center (SNFCC), including new facilities for the National Library, the National Opera and Stavros Niarchos Park. The new 22,000 m^2 NLG premises are designed to provide appropriate conditions and modern infrastructure for the storage, preservation, digitization, and use of its collections (Gkinni et al. 2017).

This paper reflects the actions and research carried out during the last 3 years in the NLG, describing the preventive conservation procedures developed prior, during and after the Library's transfer to its new premises. It also aims to highlight the collaborative approach in all actions undertaken, the strategic decision to support and to integrate Preventive Conservation as part of its collection management and the immediate and long-term benefits of this approach.

2   Collections' Transfer: A Chance to Assess and Organize

In order to prepare for the Library's transfer to its new headquarters, a "Transition Programme" was carried out between 2015 and 2018 that included five actions: 1. Collections Transition and Development, 2. Digital Services Development, 3. Public Library Department Design, 4. Audience Development and 5. Staff Training. It was well understood that the Library was not merely moving to a new building, but a new era, therefore it was its chance to advance its services and profile. The actual collections' transfer was a demanding project, well prepared in advance. It was executed in less than 4 months, from January to mid-April 2018, with the collaboration of more than a hundred people on a daily basis and 235 track drives from Vallianeio to SNFCC (SNFCC1 n.d.).

Within this context, the Conservation Lab was actively involved in actions divided into the following axes.

2.1   Collections

This category focuses on the actions applied or related to collections as physical objects. Starting with surveys, the Conservation Lab had not undertaken a widespread collection survey for a long period, apart from the conservation documentation of the manuscripts and part of the archives provided by A. Glinos[1] in the 1970's. The outcome of this survey was documented and depicted on the tailor-made cotton pouches that he provided for each object. Therefore, it was important to have a mapping of the collections' damages and their frequency per collection and per storage area. In 2016, a condition survey was undertaken for the special collections, focusing on manuscripts and archives. The survey showed that the state of Preservation mainly ranged from Fair to Good. A small percentage of objects was damaged by mold, that was not active at the time of examination and there were no signs of insect infestation. Manuscripts severely damaged by mold were isolated in anoxic conditions, sealed in ESCAL™ film slip cases using RPK oxygen scavengers and oxygen indicators (Sarris et al. 2021).

The General Collection and the Rare Books Collection were mechanically cleaned using vacuum cleaners with HEPA filters, microfibre cloths and soft brushes before transfer. This project was undertaken by external contractors, under the supervision of NLG conservators. After cleaning, both collections were thoroughly examined to detect fragile objects and to prepare valuable bindings for a non-adhesive RFID mounting. Fragile objects were placed in autoclaved bags, and volumes with detached boards were secured with cotton tape. Within the period of a few months around 80,000 items were examined and treated (Gkinni et al. 2017). Condition surveys and collections' examination provided valuable data on collections' status contributed to the objects' stabilization prior to packing, addressed specific needs for their transfer and forecasted immediate interventions at the new facilities.

2.2   Preventive Conservation

The monitoring of Temperature (T) and Relative Humidity (RH) at different venues of the Vallianeio building initiated 2 years prior to collections' transfer. Vallianeio is a historical building, designed to passively control its interior climate. Reading rooms and working areas are periodically heated or cooled to create a comfort zone for the occupants. The building's right-wing functions as a shell for a five-storey cast iron structure of the closed bookstacks. Due to its large dimensions and air capacity, the environmental conditions and fluctuations in T and RH were rarely extreme. However, all walls are exterior with no insulation. The main reading room is in the central wing, surrounded by a two-store open bookstacks. This area is periodically climatized and has a glass roof that enables natural light to penetrate. The manuscript

---

[1] Antonis Glinos (1936–1998) was professor of book conservation and head of conservation at the NLG in the 1970's.

collection was housed in locked bookstacks in the anteroom of their reading room, located on the second floor, under the roof. Similarly, this storage area has no environmental control. Moreover, doors and windows are not adequately sealed, leading to air pollutants and dust penetration and allowing the external conditions to affect the interior climate. Dataloggers for monitoring T and RH were placed in all storage areas and reading rooms and their data was manually collected. The outcome, as expected, showed that the RH remained at low levels for longer periods, while T was periodically raised.

2.3 Training of Employees

This was a critical project within the institution that the conservation team heartly supported. The institutional training program included weekly mutual-training sessions with interactive presentations by different employees and Partners, aimed to raise staff awareness and knowledge about different projects' scope and implementation. Conservators raised the significance of their actions and stressed the importance of Preventive Conservation as a collaborative approach that maintains collections' values and long-term preservation. The outcome was welcomed by all departments and paved the way for the implementation of further preventive conservation actions.

3 The New Premises of the National Library of Greece at SNFCC

SNFCC is located at Faliro Bay by the Athenian coast and is designed as a multifunctional arts, education, and entertainment complex. In 2016, it was the first large-scale cultural project to achieve LEED Platinum certification in Europe. The goal was to integrate active and passive technologies that make SNFCC one of the world's most environmentally sustainable buildings. Its energy efficiency initiatives result in a 40% energy reduction compared to what a similar building complex would consume without their implementation. All systems are designed to save energy, such as heating, air conditioning, and lighting. The planted roofs of the NLG, the National Opera and the parking buildings are covered with Mediterranean plants grown in a special substrate, creating cooling conditions for the buildings and acting as a protective layer in both winter and summer. Moreover, a large canopy is suspended above the National Opera, 47 m. above sea level. The canopy measures 10,000 m2, weighs 4,500 tons and is covered with 5,700 solar panels, able to produce up to 2,2 GWhs per year. It contributes substantially to SNFCC's energy needs and minimizes CO2 emissions (SNFCC2 n.d.).

Since its completion in 2017, SNFCC has become one of Athens' hot spots, offering a new cultural hub. One must visit the center in the summer to understand

Fig. 1 The Bookcastle at the SNFCC building. View from the central entrance of the National Library

the appeal it has for the citizens and visitors of Athens. This is a huge challenge for the NLG, as the new reality includes openness, advanced interactions with its users and the wider public, an increase in collections' demands, conservation, and digitization actions.

The SNFCC building has 3 reading rooms for the General Collection, one for the manuscripts, archives and special collections, one for the periodicals and one for the lending collection and the children's section. All reading rooms are surrounded by open bookstacks. The main storage area is located on the second floor, including closed bookstacks with different compartments, whereas the special collections are stored in five vaults. The façade of the "Book castle", comprises of four sides, two of a 20 m sided square, with a height of 18 m. that reaches the top of the building, and four levels of balconies (Fig. 1).

4 New Building New Challenges: Integrating Preventive Conservation at SNFCC Premises

In 2018, after the massive endeavor of collections' transfer, the early preservation activities focused on achieving a smooth transition to the new storage facilities. The goal was to integrate preventive conservation under a concrete approach of a collaborative scheme, that is not to undertake actions that would be isolated and mainly and exclusively operated by the conservators. Due to the time and organizational maturity limitations, the conservation team did not proceed with a strategy or policy development at the time. The plan was to follow the previous working methodology and enrich it according to demands created by the new premises. The outcome was the following three steps.

4.1 Preventive Conservation: Prioritizing Actions

Starting with identifying and documenting the strengths and weaknesses of all storage areas, the aim was to maximize the new building's potential for a safe storage environment and efficient energy use. The new facilities are tightly sealed, among others to ensure an energy-saving environment. Dataloggers for the monitoring of T and RH are placed in ten storage areas and readings rooms and one datalogger for the measurement of light (lux) and UV radiation, is placed on the façade of the Book Castle. It is a flexible and wireless monitoring system, that provides a constant view of environmental conditions.

4.2 Preventive Conservation: Assessing and Planning for Improvements

After collecting data throughout 2018, all data collected were assessed and compared to the setpoints (20 °C $\pm$ 2 °C, and 45 $\pm$ 5% RH). Although most storage areas had a rather stable environment, there was still room for improvement. Therefore, for 2019 the aim was to improve RH conditions, that is to avoid extremes values (>70% and <20%) and to create 2 zones. Zone 1 includes the cooler months of the year, approximately from November to February and zone 2 the warm months from March to October. Our goal was to keep RH within 40–50% for Zone 1 and 50–60% for Zone 2. In the case of the vaults, the data showed mild RH fluctuations throughout 2018, therefore our target for 2019 to keep the RH values within the range of 40–50% with limited fluctuations. The temperature data was rather stable, so it was decided to aim for values compatible for both humans and collections in each area, ranging from 180 to 220 °C for special collections and 200–250 °C for reading rooms. Finally, light measurements (lux and UV) were collected from the 3rd level of the Bookcastle façade. Although UV was excluded due to the filtering of the exterior glass panels, the maximum light intensity reached 12,000 lx during the winter months, that demanded immediate actions. These readings, along with blue wool standards cards, raised the importance of the situation and paved the way for a fruitful collaboration with all stakeholders. After testing different solutions and materials, it was decided to place shades to cover the Bookcastel's open bookstacks. The results were satisfactory since the light intensity varied from 0 to 600 lx throughout the day.

The BMS management and preventative maintenance is undertaken by CORDIA, a total facility management company. The BMS controls 18 central Air Handling Units with VAV or fan coils units that regulate the environment of the NLG facilities. All units have PRE-Filter (G Class-MERV 7/8) and MEDIUM Filter (F Class) F7-MERV 13. The management of BMS in order to meet preservation and human needs is a collaborative result of parties and employees with diverse expertise. Acknowledging the important of preservation, a facility management team was created with liaisons from the NLG, the SNFCC Single Member S.A.'s and CORDIA, to hold regular meetings. The impact of these meetings was obvious in the environmental data collected in 2019. The number and intensity of RH fluctuations decreased, and Zones 1 and 2 started to form. To achieve that outcome, many alterations and improvements were applied to the infrastructure. However, since 2020, the task of controlling the environment under the national regulations of providing 100% fresh air circulation put pressure on the facilities' infrastructure and the team. The target now is, within the 2 zones, to exclude the three extremes, too high, too low, too severe / frequent fluctuations.

4.3 External Collaborations

Two external collaborations with academic departments aimed at investigating the air quality in NLG premises. In collaboration with the Mycology Research Group of the Department of Biology, National and Kapodistrian University of Athens, 2 projects were launched to investigate air quality in relation to airborne fungi. In 2014 an initial project examined Vallianeio and in 2019 two more projects were executed at the SNFCC building. The projects included sampling (repeated at regular intervals) performed in indoor environments (air) in selected sites of the buildings, and the outdoor environment for comparison. The taxonomic study of airborne fungi and fungi in presumptions collected from sampling provided indications for the distribution of fungi and their possible effects, while revealing the presence of fungi that may be associated with an increased risk of cultural presumption as presented by (Pyrri et al. 2018, 2020; Iliopoulou et al. 2019; Kourteli et al. 2019).

Moreover, in 2019 the Field Analytical Chemistry and Technology team of the School of Chemical Engineering, National Technical University of Athens, carried out preliminary air quality measurements, of the installations and indoor space of the SNFCC premises by determination of the Volatile Organic Compounds in the air. Sampling was performed in 10 areas and Thermal Desorption- Gas Chromatography—Mass Spectrometry was used for the analysis. VOCs are related to the comfort zone of the employees and visitors, the health and safety of both and more broadly with security issues. Although up to now only these preliminary results are available, the goal is to integrate measurements of indoor air and relate them with the environmental and installation activities.

The outcome of the two collaborative projects enhanced awareness of the air quality status of the Library facilities and provides indexes for future monitoring. The aim is to further elaborate on these studies and to document any deterioration factors that will lead to the improvement of collections' preservation.

5 Conclusions

The last three years the NLG has undertaken a massive collections' transfer project and the relocation of its headquarters. Through numerus collections' care procedures such as surveys, objects' stabilization, storage, handling, packaging, the Library set the foundations of a strong Preventive Conservation programme, as a result to a concrete transfer project and the prospects of its new premises. The outcomes are partly attributed to the conservators' persistence and support of Preventive Conservation procedures but also the collaboration and well-intentioned communication between the stakeholders, the different departments of the NLG but also our external associates. Although organized by the Conservation Department, Preventive Conservation is a systematic and collaborative approach, not just a responsibility of the few, to ensure collections longevity. This collaborative procedure was beneficial on an organizational level, since roles and responsibilities were well defined and outlined, gaps and shortcomings were identified and addressed and the needs for future plans were spotted.

Our short-term goal is to continue building on raising awareness within the Library regarding Preventive Conservation and to proceed with forming a policy that will support continuity and consistency in the way the environment is recorded, managed and researched.

Acknowledgements I would like to thank Dr. Filippos Tsimpoglou, General Director of NLG, the conservation team, and the facility management team members from the NLG, SNFCC Single Member S.A.'s and CORDIA. Finally, all our associates from the Mycology Research Group of the Department of Biology, National and Kapodistrian University of Athens and the Field Analytical Chemistry and Technology team of the School of Chemical Engineering, National Technical University of Athens.

References

Gkinni, Zoitsa, Tsaroucha Christina, and Sarris Nikolas. 2017. The national library of Greece: Moving into a new era. In *Session 170 - preservation and conservation with rare books and special collections*, Poland, IFLA WLIC 2017 – Wrocław, Poland – Libraries. Solidarity. Society.

Iliopoulou, Stravroula, Kapsanaki-Gotsi Evangelia, and Pyrri Ioanna. 2019. Airborne fungi in closed bookstacks and collection vaults of the national library of Greece (SNFCC). In *Abstract book 16th national scientific congress of the hellenic botanical society*, 72. Athens, Hellenic Botanical Society.

Kourteli, Maria, Kapsanaki-Gotsi Evangelia, and Pyrri Ioanna. 2019. Airborne fungi in reading rooms of the national library of Greece (SNFCC). In *Abstract book 16th national scientific congress of the hellenic botanical society*, 77. Athens, Hellenic Botanical Society.

Pyrri, Ioanna, Tripyla Efstathia, Zalachori Anna, Chrysopoulou Maria, Parmakelis Aristeidis, and Kapsanaki-Gotsi Evangelia. 2020. Fungal contaminants of indoor air in the national library of Greece. *Aerobiologia* 36: 387–400.

Pyrri, Ioanna, Zalachori Anna, Tripyla Efstathia, and Kapsanaki-Gotsi Evangelia. 2018. The occurrence of airborne fungi in the national library of Greece. In *Abstract book 11th internatiol congress on aerobiology*, 70. Parma, Italy, Internatiol Association for Aerobiology.

Sarris, Nikolas, Gkinni Zoitsa, and Tsaroucha Christina. 2021. Moving the manuscripts of the national library of Greece to a new home. In *Care and conservation of manuscripts 17. Proceedings of the seventeenth international seminar held at the university of Copenhagen 11th–13th April 2018*, Charlottenlund, Museum Tuscu-lanum Press.

SNFCC1. n.d. National library of Greece. Stavros niarchos foundation cultural center. https://www.snfcc.org/en/national-library-greece. Accessed 2 January 2022.

SNFCC2. n.d. Sustainability hub. Stavros niarchos foundation cultural center. https://www.snfcc.org/en/sustainability-hub. Accessed 2 January 2022.

Enhance Performance and Reduce Energy Use in Storage Areas: Two Belgian Case Studies

Annelies Cosaert, Geert Bauwens, and Estelle De Bruyn

Abstract Resilient Storage is a project that focuses on designing a protocol for Belgian museums to implement short term energy-savings strategies, decrease CO_2 emissions and improve preservation conditions. Cultural institutions' mandate to optimally preserve their collections often encourages them to invest in Heating, Ventilation and Air-Conditioning (HVAC) systems to strictly control the indoor climate in their collection storage. These HVAC systems induce high operating costs and energy use. Today's context calls for a combined strategy, one that reduces an institution's environmental impact and at the same time sustains or even improves preservation conditions. Resilient Storage aims at helping museums to define an optimal indoor climate, taking into account the collections requirements, the capacity of the building envelope and climate control systems, and the human needs. This requires an interdisciplinary approach between preventive conservation specialists, engineers, collections and facility managers, with support from upper management. The Resilient Storage project focuses on how the building's inherent hygrothermal properties can be used in order to reduce the energy consumption for indoor climate control, while maintaining an adequate climate for the preservation of the collection. The protocols starts with a focus on documentation. The indoor climate and energy consumption of the climate control system are monitored for several months in two cultural heritage institutions to obtain insight into the buildings' current behavior. This allows identifying possible pre-existing issues, such as incorrect relative humidity, and other obstacles, such as a faulty air tightness or a detrimental ventilation rate. Based on these findings, a number of tests and interventions are defined for the case-study museums. This article will focus on several tests that are conducted in order to evaluate and improve the indoor climate at two Belgian museums: the Belgian Comic

A. Cosaert (✉) · E. De Bruyn
Sustainability Unit, Royal Institute for Cultural Heritage (KIK-IRPA), Brussels, Belgium
e-mail: annelies.cosaert@kikirpa.be

E. De Bruyn
e-mail: estelle.debruyn@kikirpa.be

G. Bauwens
Charp, Mechelen, Belgium
e-mail: geert@charp.be

203

Strip Center (Brussels, Belgium) and the FeliXart Museum (Drogenbos, Belgium). Both museums are chosen because they have different building envelopes, (HVAC) systems and collections. The FelixArt Museum houses a modern art collection. The building dates from 1996 and is equipped with a HVAC system for both exhibition rooms and storage spaces. The Belgian Comic Strip Center on the contrary is housed in a 1905 Art Nouveau building. They focus on temporary exhibitions and have therefore a smaller storage area that houses predominantly comic books. They have limited climate control consisting of one small air conditioning unit per depot. In the FeliXart museum, a first test investigates the impact of reducing the ventilation rate in the storage area. A second and third test aims at widening the acceptable bounds of both temperature and relative humidity. A fourth test investigates the impact of reducing the amount of fresh air in the ventilation air mix. At the Comic Strip Center, the first test indicated that shutting down the two air Conditioning units for several days results in a rapid, significant increase in temperature and a slow but nonetheless significant decrease in relative humidity. In subsequent tests, the plan is to reduce climate variations in the surrounding exhibition area, as this significantly impacts the storage area.

1 Context: Indoor Climate Concerns, Energy Bills and Existing Methodologies

Museums face difficult choices as they respond to apparently conflicting mandates to lower operating costs, reduce energy consumption and preserve collections. Years of research on climate and materials decay showed that temperature and relative humidity (RH) significantly impact the lifespan of collections held by cultural institutions (Michalski 2018). The indoor climate inside these facilities is thus a major concern for museums at national and international level. The desire to guarantee a stable indoor climate often encourages cultural institutions to invest in HVAC systems with high operating costs and energy consumption (between €10,000 and €55,000 per year for 900 m^2) (Image Permanence Institute 2017; Bauwens and De Bruyn 2019).

While museums are struggling to allocate funds to preserve their collections following strict guidelines, international agreements advocate for loosening climate in order towards sustainable objectives:

- The 2014 Environmental Guidelines: ICOM-CC and IIC Declaration from the International Council of Museums—Committee for Conservation (ICOM-CC) and International Institute for Conservation of Historic and Artistic Works (ICOM-CC Melbourne)
- The 2015 Bizot Green Protocol of the National Museum Directors' Council (NMDC) (Bizot Group 2015)

- The 2019 guidelines presented in Chapter 24 of the American Society of Heating, Refrigerating and Air-Conditioning Engineers (ASHRAE) (American Society of Heating, Refrigerating and Air-Conditioning Engineers [ASHRAE] 2019)

However, all these guidelines and initiatives struggle to be translated within the collection's management practices, for reasons that have been widely discussed in publications (e.g. the preference for quantifiable versus non-quantifiable information, honesty about the existing climate versus peers, 'best' practices that reinforce the importance of strict guidelines and international loan practices) (Ashley-Smith et al. 1994; Atkinson 2014; Henderson 2018).

Defining an optimal environment for the preservation of collections requires different types of information. Tools striving for improvement, addressing multidisciplinary fields, have been created:

- Guide (and methodology) to: Sustainable Preservation Practices for Managing Storage Environments (from the Image Permanence Institute, IPI) (Image Permanence Institute 2012, 2017)
- Practical Guide for Sustainable Climate Control and Lighting in Museums and Galleries (from Museums & Galleries Queensland) (Museums & Galleries Queensland 2014)
- ASHRAE Handbook, Chapter 24: Museums, Galleries, Archives, and Libraries (ASHRAE) (ASHRAE 2019)
- Managing Indoor Climate Risks in Museums (from Ankersmit & Stappers) (Ankersmit and Stappers 2017)
- DEMI MORE: une approche intégrée du processus de conservation (from Kempens Landschap, provincie Noord-Brabant) (Descamps et al. 2018)
- Analyse van en Bouwstenen voor de Uitwerking van een Programma van Eisen voor Cultureel-Erfgoeddepots in Vlaanderen (from Flemish Gouvernement) (Bertels et al. 2015)

While building further on these methods Resilient Storage add a "this-is-how-you-do-it" component. In 2018, ICCROM and CCI surveyed 444 professionals from 70 countries on their needs of preventive conservation tools. The results highlighted the desire for more practical and comprehensive resources (Katrakazis and Lambert 2018). To this day, there has been no study that translates the references above to a comprehensive and practical roadmap for sustainable storage- and exhibition spaces.

2   Resilient Storage (2020–2022)

Storage spaces are unique environments because of a number of different factors:

1. While they contain the bulk (90–95% according to ICCROM) of an average collection, they do so within a relatively compact space. This means that a change in climate impacts a large part of the collection and can cause large scale damages (e.g. development of mold). On the other hand, a large number of organic objects

and packaging materials in a space reduces humidity fluctuations as the collection itself functions as a moisture a buffer (Padfield 2015). Furthermore, furniture and packaging are a secondary and tertiary envelope around the object. The climate within these envelopes, which act like a buffer, will be different to the space itself.

2. The occupation of the storage areas is relatively low which means that one can be allowed to divert from a comfort temperature (18–20 °C), assuming that personnel does not work in these spaces. In Belgium, winter temperatures are low, which means that heating is often our predominant source of energy consumption. Additionally indoor heating decreases the humidity which means that, without moisture control, humidity can drop below 30%. Choosing a colder temperature in winter can therefore be an energy friendly alternative, that discourages the development of mold and allows for better moisture control.

3. Museums can organize collection storage per material. While many 'general collections' require a similar climate, specific collection (e.g. metals and photography, prefer a colder and dryer climate then average, often referred to as 'cool' or 'cold' storage).

All these factors—and the decision-making around them—can have an big impact on energy consumption and collection related risks.

The protocol developed for Resilient Storage focusses on short-term improvement of the 'variable' factors such as the HVAC system settings: temperature and relative humidity set points, quantity of outside air entering the climate system, total air flow, and system hours of operation. The 'fixed' factors (such as outdoor climate and building envelope) will not be manipulated.

The protocol is especially developed for small-scale museums, which have to deal with a limited amount of financial and human resources. Two complementary case studies are chosen to fit the scope of this project: The FeliXart Museum and The Belgian Comic Strip Center and the FeliXart Museum.

3 Pilot Projects

Both pilots present very different opportunities to improve the collection environment, and save energy. For each pilot, we highlight and argue for the different tests. Among others, these tests consider the response time of objects and the schedule of in-house collaborators. Before the implementation of tests, the project team ensures all information is at hand to appropriately evaluate the impact of the implemented measures. During the tests, they regularly assess onsite conditions do not exceed pre-defined thresholds. After test completion, they evaluate the test against the anticipated results. Based on these evaluations, measures that are shown to perform well can be implemented.

3.1 Belgian Comic Strip Center (Brussel, Belgium)

3.1.1 Description

The Belgian Comic Strip Centre (1989) is a private non-profit organization that houses a collection of more than 10,000 original drawings. Located in an Art Nouveau-style building designed by renowned architect Victor Horta in 1906, it was initially intended to accommodate a textile warehouse. Hence its architecture is not the most appropriate for exhibition and preservation of graphic works of art on paper. The building has large open spaces that unfold on several levels around a central glass roof. The building is not insulated. The building is heated with a central gas-powered heating system, and the individual storage areas are conditioned with air-conditioning units. The museum's annual energy costs in 2018 amounted to about 14% of the total budget (including staff costs).

The two storage areas at the museum measure just 15 square meters each (Fig. 1). Collections in storage are mainly graphic works of art on paper. These are stored in archival boxes, or in drawing cabinets with interleaf paper acting as buffering and protective material. The air conditioners' set-point temperatures are set to 18 °C. They perform no active moisture control.

3.1.2 Historical Data, Observations and Tests

Due to a low insulation quality and high air permeability of the outdoor fabric, the indoor climate in the exposition rooms is significantly correlated with the outdoor climate. A first test was designed to look at the impact these exposition rooms have on the storage areas. During two three-day periods in April 2021, the air conditioning unit in storage area 2 was switched off (Fig. 2). The temperature data seems to indicate that the airtightness between the storage area and the surrounding exposition rooms is flawed. Additional airtightness tests are needed to verify that. However, the impact on relative humidity levels remains very limited. As the storage room has limited hygroscopic material, the archival materials are likely to add a significant moisture buffer capacity.

Previous research indicated that during summer, indoor temperatures can rise to levels exceeding 30 °C. As the building is opened during the day and closed off during the night, it is continuously 'charged', resulting in ever higher temperatures over the course of warm summer periods. A test was envisaged to limit this summer overheating by means of night ventilation (Fig. 3) Because of security reasons this test cannot be performed. However, in the framework of a larger-scale renovation works, secure ventilation in the glass roof structure will be considered.

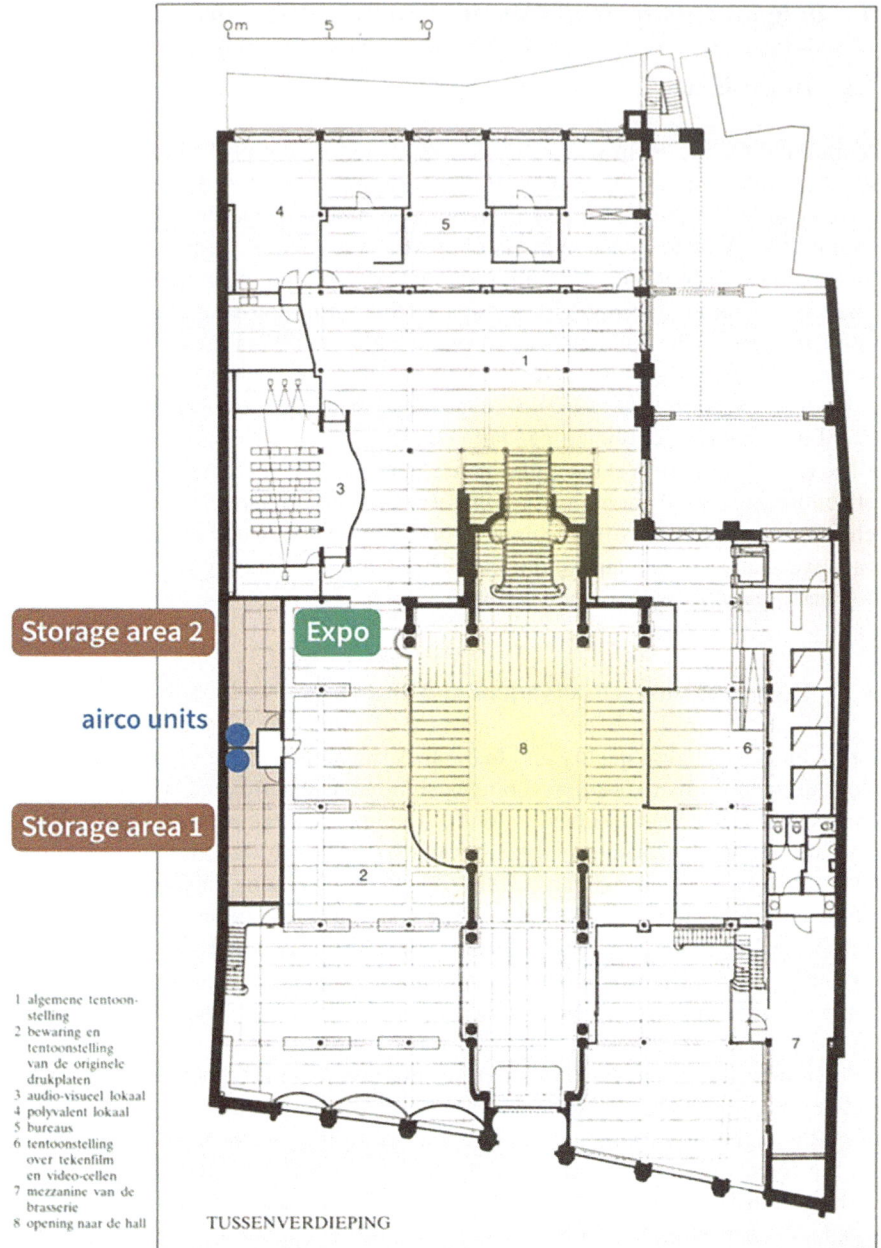

Fig. 1 Mezzanine floor, with exposition rooms ("Expo") organised around a central space. Two small storage areas are situated adjacent to the party wall

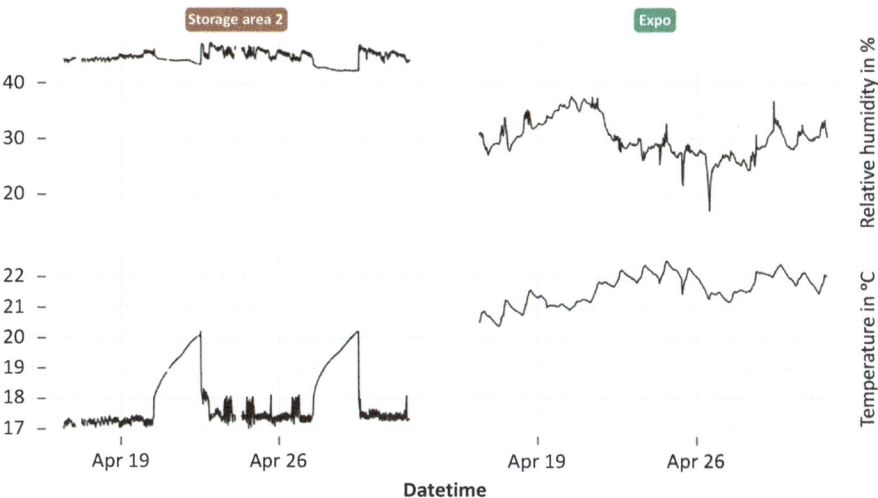

Fig. 2 Relative humidity (top) and temperature (bottom) in Storage area 2 (left) and the adjacent exposition room "Expo" (right) during repeated three-day air conditioning shutdowns

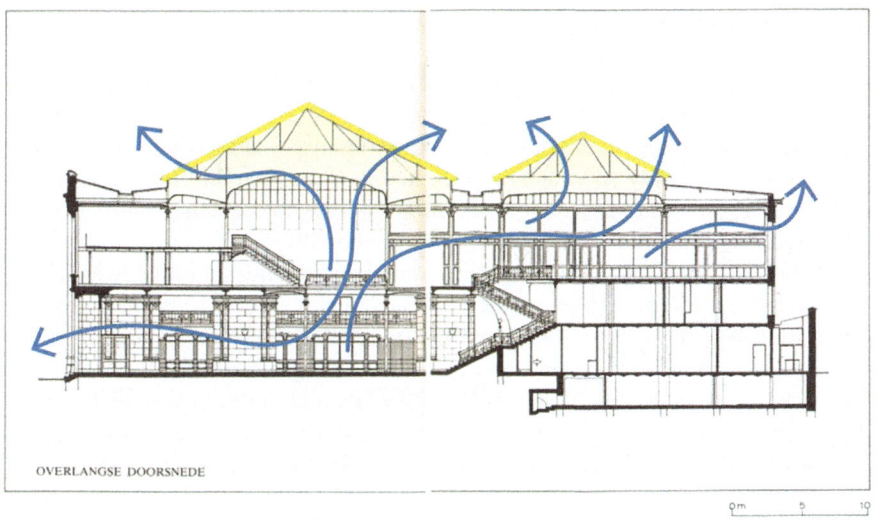

Fig. 3 Section illustrating the building entrance and glass roofs that might be operated to support night ventilation

3.2   The FeliXart Museum (Drogenbos, Belgium)

3.2.1   Description

The FeliXart Museum (1996) is situated in Drogenbos, Belgium, on domain of 5 hectare that once belonged to Felix De Boeck (1895–1995), one of the pioneers of Belgian abstract art. The museum showcases historical and contemporary avantgarde art at the museum and involves the local community through activities in the domain surrounding the museum. The FeliXart Museum is overseen by a nonprofit organisation created and financed by three government bodies. The museum is located in a modern building erected in 1996, with insulated walls and roofs, and double-glazed windows. Since 2012, the museum's indoor climate is conditioned by a complete HVAC system with winter comfort heating and humidification, and summer cooling and reheating for dehumidification. The settings are very strict: a temperature of 20 °C and a relative humidity of 50% ($\pm2\%$ daily variation, and ±1 °C seasonal drift). Annual energy costs amount to about 12% of the museum's total budget (including staff costs).

3.2.2   Historical Data, Observations and Tests

Current ventilation flow rates for the storage areas are much higher than required for hygienic ventilation and climate conditioning. From the outset, they were based on office spaces. But storage areas have a very different use. For instance, storage areas 1 and 2 (Fig. 4) are only sporadically occupied (a short period of time one day per week). A first test will thus reduce the ventilation flow rate for these storage areas, thereby reducing the outdoor air that needs to be conditioned.

A second test will look at relaxing temperature and relative humidity levels. From the available historic data (Fig. 5), there already seems to be a seasonal setting for relative humidity in storage area 1, going from 40% in winter to about 60% in summer. The current settings can be widened, without going outside the boundaries of ASHRAE's 'climate class A' (ASHRAE 2019), by relaxing temperature. Reducing indoor air temperature during winter is likely to reduce humidification demand.

A third test will decrease the ratio of outdoor to recirculated air. This test is similar to the first, as both serve to reduce the amount of fresh outdoor air that needs to be conditioned. Currently, the supply air comprises 90% fresh air and 10% recirculated air.

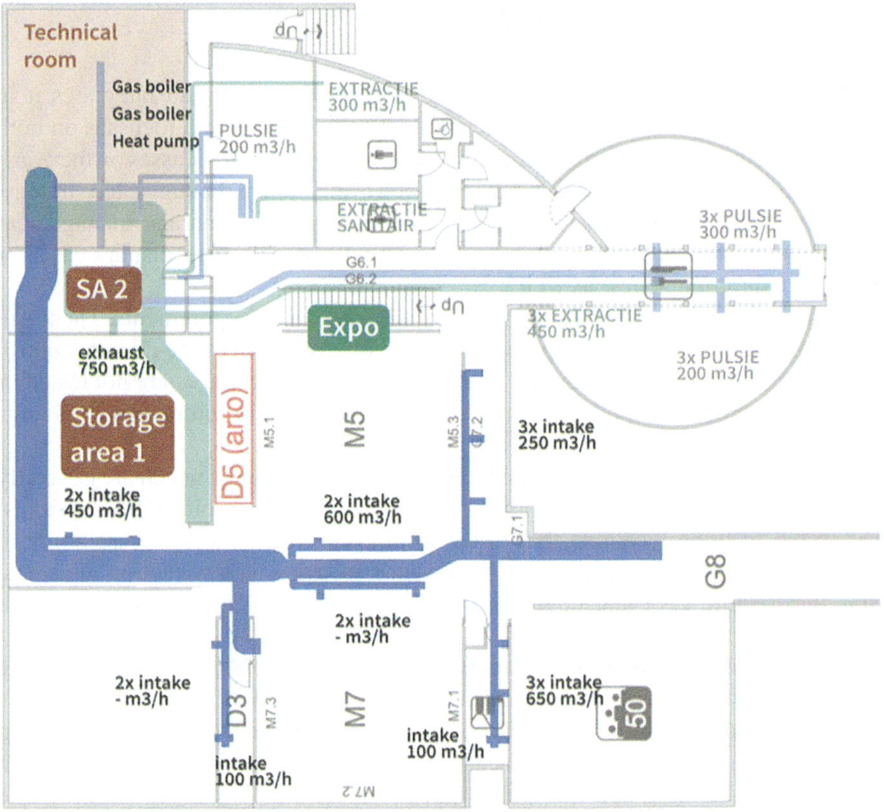

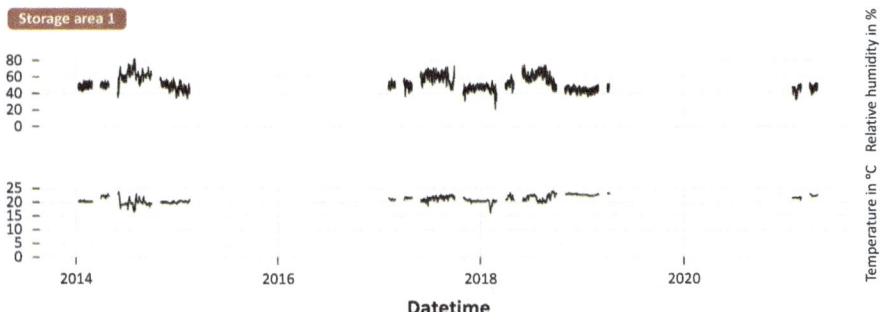

Fig. 4 Floor plan FeliXart museum. A shared air handler serves to condition the air in storage area 1 and 2 and the *downstream* exposition rooms

Fig. 5 Historic relative humidity (top) and temperature (bottom) in Storage area 1. Temperature levels remains rather constant around or slightly above 20 °C. Relative humidity levels vary between 40% in winter and 60% in summer

4   Climate2Preserv (2021–2025)

C2P takes the knowledge gathered through Resilient Storage and integrates it in
a more extensive protocol that considers all collection spaces and focusses on both
long- and/or short-term improvements. The short-term protocol focusses on the 'variable' factors and includes an implementation and evaluation of the suggested strategy.
However, it includes possible changes in collection presentation and management
(e.g. changing visitor flow, working with microclimates such as display cases or
changing exhibition zones) as well.

The long-term protocol ends with the formulation of new requirements. These can
be changes to the building envelope, renewal or improvement of the climate control
system, as well as a move of the collection. The suggested changes are not tested or
evaluated within the scope of this project, but a production of future energy savings
will be calculated.

C2P is firstly developed for the Federal Scientific Institutions, which are generally
located in historic buildings with large-scale collections and complex HVAC systems.
It is tested in three locations that differ from each other. The first case study at
CINEMATEK (Royal Belgian Film Archives) is focusing on a collection of acetate
films that require cool and cold storage. This project will implement short-term
improvements.

At the Royal Museum for Fine Arts Brussels (KMSKB-MRBAB), one case study
is focusing on short-term improvements to a complex HVAC system in recently
renovated museum exhibition spaces, conservation areas and storage spaces. During
a third case-study the protocol is tested aiming at formulating a long-term transition
of a historical building and exhibition space at Museum Wiertz (Fig. 3, a satellite
museum of KMSKB-MRBAB) (Climate2Preserv 2020).

5   Conclusion: Towards Sustainable Cultural Institutions
in Belgium

For Resilient storage and C2P the "how-do-you-do-it" component is essential. There
is a focus on learning from each other, working as a team, creating a common
understanding and striving towards a clearly defined goal. However, every team has
different capacities and every institution works within a different reality (budget,
staff, collection, public/private, etc.). It is therefore extremely important that the
protocols created are flexible. We therefore strive to reuse or refer to existing sources,
recognizing others efforts, and to have a modular protocol, that allows you to explore
different paths rather than requiring you to follow it from beginning to end.

Small to mid-size organisations, often have a lot of 'in house' knowledge. But
most of them will require outside expertise to complete the full protocol. This outside
expertise can either focus on management, collection, technical aspects or all of the
above, based on the organisations needs. However, the 'in house' knowledge of the

employees remains key and without active collaboration with several departments and support from upper management, the project cannot succeed.

Because some of the structures of these institutions are complex (different institutions are responsible for building, system, collection, daily operations, maintenance, etc.), it seem opportune to work with 'framework agreements' in the future. These agreements will allow the outside experts to buy and install certain materials and avoid discussion about allocating funds/time/personnel, minimize admin and therefore speed up the process.

With Resilient Storage and C2P, we want to align the research potential to the societal needs in terms of response to climate change, answering the worldwide pressure on public institutions and society for more sustainable practices in every sector (such as recalled by the recent COP 26 UN Climate Change Conference). The Belgian cultural sector can play an exceptional role here: it has an excellent cultural and scientific in-house capacity to collaborate on, and pioneer the implementation of nation-wide methodologies.

Acknowledgements The authors would like to warmly thank the partners and funding bodies of Resilient Storage and Climate2Preserv.

Resilient Storage: FARO/Vlaams steunpunt voor cultureel erfgoed, Musées et Société en Wallonie (MSW), urban.brussels, the Belgian Comic Strip Center, the FeliXart Museum, ICOM Belgique Wallonie-Bruxelles, ICOM Belgium Flanders. Funding bodies: the Brussels-Capital Region (urban.brussels), the Flemish Government (Department Youth, Culture and Media) and the Fédération-Wallonie Bruxelles.

Climate2Preserv: KU Leuven, University of Liège, Royal Museums of Fine Arts of Belgium, CINEMATEK, ICCROM, Academia Belgica. Funding body: the Belgian Science Policy Office (Belspo).

References

American Society of Heating, Refrigerating and Air-Conditioning Engineers (ASHRAE). 2019. *Museums, galleries, archives, and libraries: ASHRAE handbook*. Atlanta: ASHRAE.
Ankersmit, Bart, and Marc H.L. Stappers. 2017. *Managing indoor climate risks in museums*. New York: Springer.
Ashley-Smith, Jonathan, Nick Umney, and David Ford. 1994. Let's be honest: Realistic environmental parameters for loaned objects. In *Preventive conservation practice, theory and research*, 28–31. London: The International Institute for Conservation of Historic and Artistic Works.
Atkinson, Jo K. 2014. Environmental conditions for the safeguarding of collections: A background to the current debate on the control of relative humidity and temperature. *Studies in Conservation* 59: 205–212.
Bauwens, G., and E. De Bruyn. 2019. *Survey on the energy consumption of Belgian museums for the research project Resilient Storage*. Belgium: KU Leuven & KIK-IRPA.
Bertels, Inge, Dorien Aerts, and Filip Descamps. 2015. *Cultureel – erfgoeddepots. Analyse van en bouwstenen voor de uitwerking van een programma van eisen voor cultureel-erfgoeddepots in Vlaanderen. Onderzoekopdracht*. Brussels: Departement Cultuur, Jeugd, Sport en Media.
Bizot Group. 2015. Bizot green protocol. In *Environmental sustainability: Reducing museums' carbon footprint*. London: National Museum Directors' Council.

Climate2Preserv. 2020. Sustainable climate management strategy to preserve federal collections. BRAIN-be 2.0 (technical fiche). https://www.belspo.be/belspo/brain2-be/projects/Climate2Preserv_E.pdf. Accessed 25 November 2021.

Descamps, Filip, Dries Haesendonck, Philippe Lemineur, Griet Bronselaer, and Roald Hayen. 2018. *DEMI MORE: Une approche intégrée du processus de conservation.* Belgium: Interreg Vlaanderen-Nederland.

Henderson, Jane. 2018. Reflections on the psychological basis for suboptimal environmental practices in conservation. *Journal of the Institute of Conservation* 41: 32–45.

ICOM-CC Melbourne September 15–19, 2014 and IIC Hong Kong September 22–26, 2014.

Image Permanence Institute. 2012. *IPI's guide to: Sustainable preservation practices for managing storage environments.* Rochester: Image Permanence Institute.

Image Permanence Institute. 2017. *IPI's methodology for implementing sustainable energy-saving strategies for collections environments.* Rochester: Image Permanence Institute.

Katrakazis, T., and S. Lambert. 2018. *Outils et ressources en conservation préventive: Point de vue des utilisateurs.* Rome: ICCROM.

Michalski, Stefan. 2018. *Agent of deterioration: Incorrect temperature.* Ottawa: Canadian Conservation Institute.

Museums & Galleries Queensland. 2014. *A practical guide for sustainable climate control and lighting in museums and galleries.* Sydney: International Conservation Services and Steensen Varming.

Padfield, Tim. 2015. Calculation of the energy used for dehumidification. https://www.padfield.org/tim/cfys/atmcalc/dehumidcalc-help.html. Accessed 19 November 2021.

Diverse Collections, Different Contexts: Risk Management for the Oswaldo Cruz Foundation's Cultural Heritage

Marcos José de Araújo Pinheiro⊙, **Carla Maria Teixeira Coelho**⊙, and **Ana Roberta Tartaglia**⊙

Abstract The article aims to reflect on the importance of implementing risk management for the conservation of diverse collections in different contexts. The discussion is based on the experience of teams at the Oswaldo Cruz Foundation in Rio de Janeiro, Brazil, a century-old institution devoted to research and development in public health and that has various types of historical and scientific collections in its custody. From 2014 to 2018, Casa de Oswaldo Cruz, the unit in charge of the foundation's heritage, conducted a pilot experience in the implementation of risk management for the Fiocruz collections, adopting the ABC Method developed by the International Centre for the Study of the Preservation and Restoration of Cultural Property (ICCROM) and the Canadian Conservation Institute (CCI). The result of this pilot experience was essential for expanding knowledge of the different risks that can impact the foundation's cultural heritage and for proposing mitigation measures. The COVID-19 pandemic brought new challenges for the institution and required rethinking strategies to guarantee the safety and security of collections, employees, and users, considering contextual changes that included alterations in the uses of the sites where the collections are located and the ways of accessing them.

1 Introduction

Risk management represents an important paradigm shift in the field of cultural heritage. Authors such as Brokerhof (2006) and Michalski (2016) evaluate that for many years preventive conservation was based mainly on monitoring and environmental control. Risk management created the possibility of a more comprehensive

M. J. de Araújo Pinheiro (✉) · C. M. Teixeira Coelho · A. R. Tartaglia
Casa de Oswaldo Cruz/Oswaldo Cruz Foundation, Rio de Janeiro, Brazil
e-mail: marcos.pinheiro@fiocruz.br

C. M. Teixeira Coelho
e-mail: carla.coelho@fiocruz.br

A. R. Tartaglia
e-mail: ana.tartaglia@fiocruz.br

Á. F. Perles-Ivars et al. (eds.), *Collection Care*, Springer Proceedings in Archaeology and Heritage, https://doi.org/10.1007/978-3-031-85655-6_18

view of the different threats to cultural property, allowing the identification of risks and prioritization of mitigation measures, considering complex and varied sets of cultural property based on interaction between different stakeholders. As stated by Brokerhof (2006, p. 5), "Collection risk management is not only where all our scientific knowledge and understanding of the materials in our collections comes together; it is also where the entire cultural heritage profession comes together."

The current study reflects on the importance of implementing risk management for the conservation of diverse collections in different contexts. The discussion is based on the experience of teams at the Oswaldo Cruz Foundation (Fiocruz) in Rio de Janeiro, Brazil, a century-old institution devoted to research and development in public health and that has various types of cultural property in its custody: documental, bibliographic, museum, and biological collections, historical buildings, and archaeological sites (Fig. 1). Affiliated with the Brazilian Ministry of Health, Fiocruz was founded in 1900 to produce antisera and vaccines to combat the plague epidemic impacting the Brazilian population. Over time, the institution's scope of work was expanded considerably, and various collections were created as the result of its work. In addition to the institutional recognition of its value and importance for society, part of this heritage has been formally acknowledged by the Brazilian National Institute of Historical and Artistic Heritage (IPHAN) and the UNESCO Memory of the World Program (MOW).

Aimed at establishing common parameters and guidelines for the preservation of its cultural heritage, Fiocruz has developed its preservation policy and various reference documents with preventive conservation and risk management as the underlying premises. An institutional network—the Fiocruz Collection Complex (Preservo)—was created to connect the various units responsible for managing the institution's

Fig. 1 View of the campus, Fiocruz headquarters, Rio de Janeiro. *Source* Fiocruz Collection. Photograph: Peter Ilicievv

collections across Brazil. This experience has strengthened the institution's expanded notion of cultural heritage as a strategic field for its work, the dialogue between different stakeholders responsible for the collections' preservation, and the development of projects based on preventive conservation and risk management. The experience has also proven essential for expanding knowledge on the different risks to the institution's cultural heritage, especially unusual situations and challenges such as those posed by the COVID-19 pandemic or that can result from transformative phenomena in everyday routines, and thus the possibility to review planning, strategies, and procedures and to respond quickly and assertively to mitigate risks to the cultural property and the personnel involved.

2 Pilot Cycle in Risk Management Implementation

Fiocruz entrusted one of its institutes, Casa de Oswaldo Cruz (COC), to propose a system to link and integrate the foundation's various types of cultural collections under its custody. The system, known as Preservo, began in 2008 based on work that encompassed multi-user platforms and the complex systems concept, in which the parts interact with each other, preserving their characteristics but shaping something with specific emerging properties that extrapolate the sum of the constituent elements. This approach was essential for the implementation of Preservo and its full adoption by the entire Fiocruz, given that the system preserves the autonomy of preservation activities and access by its institutes and respects the diversity of types of collections and their respective cultures and practices, but under the aegis of integration and synergy. Preservo has always operated as a network of the institutes responsible for the collections to establish integrated management, acting in an advisory capacity and formulating actions in preservation and access. The system's institutionalization occurred in 2018 with the approval and publication of the Preservation Policy for the Scientific and Cultural Collections of Fiocruz (Fundação Oswaldo Cruz 2020). Preservo assumed the responsibility for linking and orienting the policy's implementation, acting through a framework that encompasses four dimensions: conceptual; normative and referential; preservation and physical access; and preservation and digital access. The current work featured the first two, whose effects unfold in the other latter two dimensions. The conceptual dimension involves values, principles, and underlying orientation, including preventive conservation and integrated management. The normative and referential dimension aims at the formulation and publication of references and standards for preservation activities and access to collections and includes policies, plans, programs, and specialized manuals, highlighting not only the Fiocruz preservation policy but also the previous policy, published in 2013, pertaining to the collections under the custody of COC (Fundação Oswaldo Cruz 2020; Casa de Oswaldo Cruz 2013).

This set of initiatives made it possible to carry out a pilot experience with the implementation of risk management for the institution's collections, adopting for this purpose the ABC Method developed by the Canadian Conservation Institute

(CCI) and the International Centre for the Study of the Preservation and Restoration of Cultural Property (ICCROM) (Michalski and Pedersoli Jr 2016). After defining the object of this first phase, a multidisciplinary Working Group was established with experts at COC responsible for the conservation of various types of collections and representatives from management areas, besides collaboration by professionals from other institutes of Fiocruz and consultancy by José Luiz Pedersoli Jr. The ABC Method, in keeping with ISO 31000 (ABNT 2018), proposes specific tools for cultural heritage, such as layers of the collections' enclosure, and ABC scales for risk analysis related to 10 agents of deterioration (physical forces; criminal acts; fire; water; pests; pollutants; light/UV radiation; incorrect temperature; incorrect relative humidity; and dissociation). The procedural stages include the following: establishment of context, identification of risks, risk analysis, risk evaluation, and treatment of risks, besides on-going actions in communication and monitoring.

Risks are identified for each of the agents of deterioration based on the different layers of the collection's enclosure (from packaging to region), classified according to the ABC scales as to the possibilities of an incident occurring in a specified period, which defines component A, followed by analysis of the risk's impact on each item affected in the collection, component B, and the portion of the collection that may suffer from the risk and the resulting loss of value, resulting in component C. The sum of these three components results in the Magnitude of Risk (MR), which then orients the subsequent stages of assessment and the dynamics in the prioritization of risk treatments.

An excellent example for understanding these two stages is the Moorish Pavilion, the Fiocruz headquarters, listed as a national heritage building by the Brazilian National Institute for Historical and Artistic Heritage (IPHAN). For this cycle, and especially for this historical building, it was evident that the greatest risk was related to the "fire" category, reaching 12 on a scale of 15, thus characterized as an extreme priority and clearly standing out in relation to the other risks, especially due to the characteristic logarithmic progression in the MR calculation (Fig. 2). The application of the ABC Method in the Fiocruz pilot cycle resulted in the prioritization of measures to mitigate fire hazards in the various historical buildings and movable collections and has unfolded in on-going action by the Working Group in other cycles involving the methodology's application.

The approach adopted in this cycle included the following: three historical buildings (Moorish Pavilion, Stable, and Clock Pavilion); the permanent archival collection of Fiocruz; the collection of the Library of the History of Sciences and Health (BHCS); and the museum collection. Unlike the studies adopted for the architectural heritage and museum collection, which included comprehensive analyses of their context, history, use, and possible threats, the same did not occur with the archival collections and the BHCS, since they were stored at the Campus Expansion and were scheduled to be transferred to the Center for Documentation and History of Health (CDHS), still under construction at the time on the Manguinhos Campus (the main Fiocruz campus). The survey of these collections thus focused on the move to their new storage area, as observed in Fig. 2, with the risks identified and prioritized based on MR. In the article's final section, we will address the transfer of these collections,

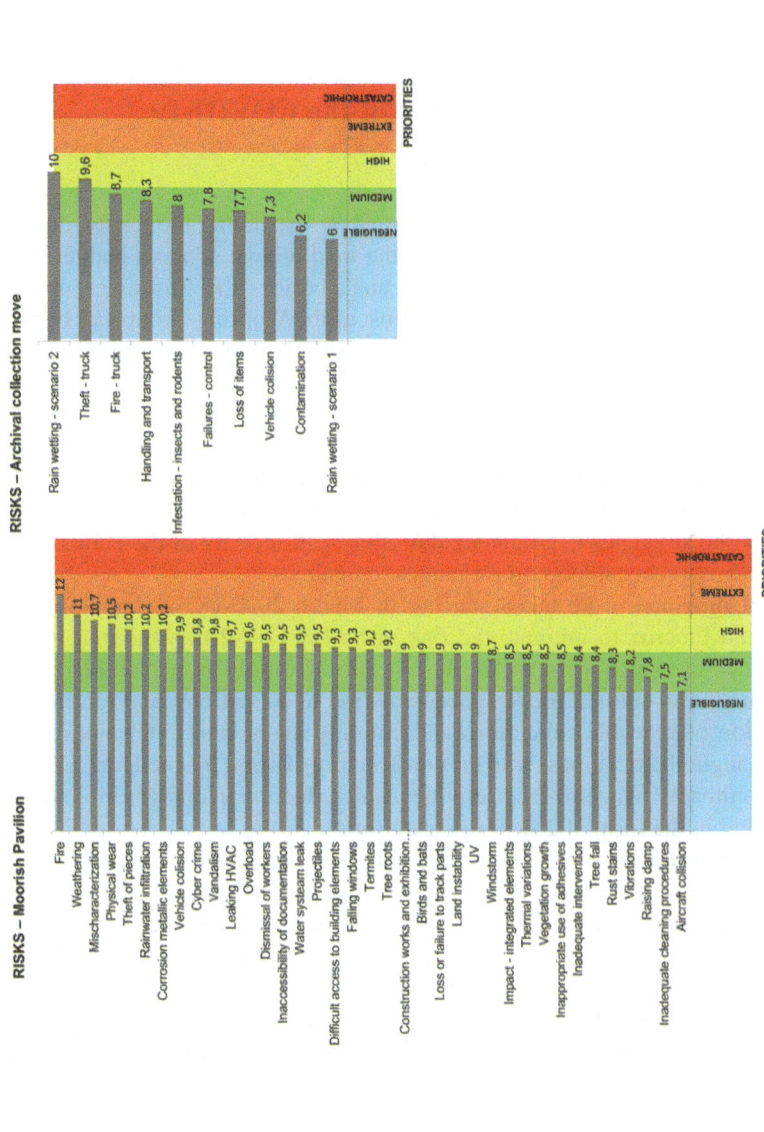

Fig. 2 Risk assessment graphs for the Moorish Pavilion and archival collection during the move, in decreasing order of magnitude (MR). Prepared by José Luiz Pedersoli Jr. ©Fiocruz. Fundação Oswaldo Cruz and Casa de Oswaldo Cruz (2020, p. 247)

since the strategies and planning required a revision to adapt to the new challenges posed for its implementation during the COVID-19 pandemic.

The results of this pilot experience were published (Fundação Oswaldo Cruz and Casa de Oswaldo Cruz 2020) and made available in ARCA, the Fiocruz's institutional open-access repository.

3 Risk Management in the Context of the COVID-19 Pandemic

The pilot cycle of risk management was essential for expanding the teams' understanding of the collections' vulnerability and the institutional context, budget, and characteristics of the sites where they are located, besides the various groups of stakeholders that interact with the collections. Just as the Working Group was beginning the second cycle, the COVID-19 pandemic struck, and it was necessary to replan the activities. The pandemic brought new challenges for Fiocruz, acknowledged by the World Health Organization (WHO) as a reference in coronavirus for the Americas. Such challenges have included proposing actions for dealing with the pandemic that include patient care and production of diagnostic tests and vaccines for the new coronavirus. In relation to the institution's cultural heritage, it was necessary to rethink the strategies to guarantee the safety and security of the collections, employees, and users, considering the changes in context.

A series of measures were taken starting in March 2020 by various levels of government and Fiocruz to mitigate the risk of SARS-CoV-2 transmission in the population. Such measures included the suspension of various in-person activities, including visitation to cultural property, consultation of collections, and routines in the conservation of movable collections and building maintenance. Changes were made in the Manguinhos Campus (with the construction of a new hospital for COVID-19 patients, a cutback in face-to-face activities, and reduction in conservation teams in the field), and in the form of access to collections.

This new context has required a new approach to emergencies that includes three stages: Stage 1 relates to all the necessary preparation for closing the institution to the public; Stage 2 covers the period in which the institution remained closed to the public and with reduced in-person activities; and Stage 3 relates to the process of resuming in-person activities. The methodology for this moment was proposed by ICCROM, providing a set of documents on its site about "Heritage in Times of COVID". The Working Group also had the opportunity to contribute by translating part of the materials provided by ICCROM into Portuguese.

The decrease in in-person activities may exacerbate some risks such as theft, fire, flooding, biodeterioration, and vandalism. In fact, various incidents have occurred during the pandemic, such as the fire in the Museum of Natural History at the Federal University of Minas Gerais (Brazil), the theft of a Van Gogh painting from the Singer Laren Museum (Netherlands), and the flood in the HERstory Museum (USA). When in-person activities are reduced in institutions, it is not possible to rely on the teams' routine watchful eye identifying risks and acting to avoid or interrupt a problem when identified. If an incident occurs with the cultural property, the response time may be much longer than usual.

Stage 1 thus covered a risk analysis considering this change in context brought by the pandemic. Although Fiocruz had already undergone a risk analysis cycle, it was important to update the information considering the risks that could be exacerbated by the situation and the identification of new risks—especially related to the health of staff and visitors.[1] As in any risk management process, it was necessary to analyze scenarios, making projections of the pandemic's impacts on the local community, the vicinity of the institution, the institution itself, and the teams.

The preparation for shutdown needed to consider all the possible measures to mitigate the risks as identified. Working with scenarios facilitated the decision-making process. In relation to the climate control, for example, it was necessary to assess each collection's situation to determine whether the HVAC equipment would remain turned on. This decision needed to consider the collection's vulnerability and the existing type of climatization equipment. In areas where the collections' climatization was done by window-mounted or split air conditioning equipment, without automation, it was necessary to turn off the equipment due to the possible impacts from malfunctioning of the equipment, potentially leading to a fire.

In Stage 2, a minimum staff contingent was determined (safeguarding persons in risk groups and considering the use of personal protective equipment—PPE and hygiene measures recommended by the Ministry of Health) for teams working in the conservation of collections and buildings, as well as for cleaning personnel. These teams began to work in a shift system to perform emergency services, as well as periodic inspections and verification measures especially in the electrical installations, cleaning gutters and rainspouts, and periodic cleaning of the places where the collections are kept, under the supervision of technical teams, aimed at minimizing risks. A protocol was determined for periodic monitoring of areas where collections are stored to collect data on the rooms' temperature and relative humidity and to verify the potential presence of insects, mold, and other problems.

Throughout the process, the Working Group members aimed to identify references to research, protocols, and scientific articles related to the pandemic's impacts,[2] collecting and analyzing references and sharing information on the topic with other Brazilian institutions. For definition of the guidelines in Stage 3, which began

---

[1] For this stage, the group used the forms developed by ICCROM—Initial Rapid Assessment Template for Identifying Risks, Monitoring Impacts, Assessing Needs for Movable Cultural Heritage and Initial Rapid Assessment Template for Identifying Risks, Monitoring Impacts, Assessing Needs for Movable Cultural Heritage.

[2] For example, the result of studies on the persistence of SARS-CoV-2 in the materials comprising the collections, according to the REALM Project—Reopening Archives, Libraries and Museums, coordinated by the Online Computer Library Center, Inc. (OCLC), the Institute of Museum and Library Services, and Battelle. We also highlight the materials produced by such institutions as the International Council of Museums (ICOM) (ICOM-BR 2020) and the Brazilian Institute of Museums (IBRAM) (Ibram 2020).

in November 2021, these references were revisited and updated according to the evolution in the available information on risk strategies to mitigate COVID-19 transmission.

Key parameters were established for the safe return to in-person activities, including the acquisition of PPE according to the different activities performed by the teams (conservation of collections, services for visitors and users, etc.) and the adoption of cleaning and disinfection protocols. Considering that the recommended materials for disinfection generally impact the materials comprising the collections, the solution found was to quarantine the items consulted or manipulated by the teams (3–5 days). This stage also included the adaptation of workspaces, consultation of collections, and exhibit areas, based on safety guidelines. This required reviewing the capacity for the public in the spaces, based on the need for physical distancing of at least 1.5 m and reassessing the exhibits' circuits, prioritizing one-way circulation and reducing the surface area potentially touched by the public to minimize the possibility of contamination.

The guidelines produced by the Working Group were incorporated into the *Plano de Mitigação dos riscos de transmissão da Covid-19 no ambiente de trabalho da Casa de Oswaldo Cruz* (FIOCRUZ/COC 2021). A communication plan was developed and revised over the course of the various stages to inform employees, collaborators, and the public on the safety measures and the new rules for consultation and visitation.

4 Moving the Collections

One of the challenges during the pandemic was transferring the archival and bibliographic collections to a new facility, especially built to house them, the CDHS. Preparations for this move began even before the building was finished, based on data from pilot cycle of risk management 2014.

The archival and bibliographic collections had been stored in the Campus Expansion (so-called because it is an extension of the Manguinhos Campus) since the creation of COC 35 years ago. The Department of Archives and Documentation (DAD), responsible for the custody of the permanent archival collection of Fiocruz, and the Library of the History of Sciences and Health (BHCS), responsible for the bibliographic collection, adapted the building's original environments to house their collections safely and securely. Thus, all the storage areas have been climatised with conventional air conditioning and dehumidifiers, indispensable items for indoor climate control in a city like Rio de Janeiro which easily exceeds 90% relative humidity, especially in the summer, and an inert gas fire suppression system.

The Campus Expansion is located several hundred meters away from the Manguinhos Campus on the opposite side of the city's main thoroughfare, Brazil Avenue (Fig. 3), and houses various units of Fiocruz with different lines of work, subject to constant overload and energy and internet outages. These conditions add

Fig. 3 View of the Campus Expansion and the building that housed the archival and bibliographic collections (top); path for the move to the Manguinhos Campus and CDHS (bottom). Photos: Google Maps, Ana Roberta Tartaglia and Silmara Mansur

to others such as workforce insecurity and proximity to the heavily traveled thoroughfare, leading to excess dust, and pollutants, impacting the collections stored there.

With the construction of the CDHS, COC achieved the double objective of creating specialized storage areas for housing the archival and bibliographic collections under technically recommended conditions according to national and international standards and the ability to group in a single building diverse sectors of the unit that were scattered across different spaces. The construction posed several challenges such as adaptation of the architectural project following archeological findings on the plot of land (the vestiges of one of the first furnaces for incinerating urban waste, used from the nineteenth to the early twentieth century). The new building has 3652 square meters of floor space, with five floors and six storage units for the collections, divided as follows: two for the textual archival collection, two for the bibliographic collection, one for the iconographic collection, and one for the audiovisual and sound collection. Criteria were adopted for environmental sustainability, contributing to the institution's sustainable development policy. The CDHS was the first public-service building in Brazil to receive maximum energy efficiency classification from the National Program for Energy Efficiency in Buildings (Procel Edifica).

Concluded in 2018, the building was occupied gradually, and only the most complex part of the archival and bibliographic collections remained to be moved, postponed due to a revision in 2019 in the HVAC system for the storage areas to guarantee the temperature and relative humidity specifications in these spaces. The move was replanned for 2020 and was postponed again due to the COVID-19 pandemic, but it was not possible to delay it longer, despite the health concerns, since new problems added to the existing ones.

The archival and bibliographic collections in their areas in the Campus Expansion had remained without climatization since Fiocruz adhered to the contingency measures against the spread of COVID-19 in early March 2020, given the need to permanently disconnect the HVAC equipments that was posing a potential fire hazard to the collections. All the areas with collections belonging to the COC—library, archives, storage area of the Museum of Life, and the historical buildings of the Heritage Site, although closed to the public and to consultations, continued to have their conservation conditions monitored periodically by professionals in each of these areas. In the case of the archives and library, we used wi-fi dataloggers to upload data via the cloud and for remote follow-up.

Although the new storage areas were ready for use in the CDHS, the beginning of the move was premised on identification of the proper timing, with a decrease in COVID-19 cases and immunization of the teams involved. With the beginning of workers' vaccination (with the COVID-19 vaccine manufactured by Fiocruz itself under an agreement with the AstraZeneca biopharmaceutical company and the University of Oxford), it was possible to plan the move for 2021 (Fig. 4). Complete vaccination was a condition for conducting the move starting in August 2021, when most of the workers were fully immunized.

The research done in the pilot cycle of risk management indicated guidelines for mitigation of risks such as rainfall and cargo theft during the move, and additional measures were taken to decrease the risk of COVID-19 transmission. The move was

Fig. 4 Scenes of the move: textual archives and bound volumes in their new storage area (A and B); Arrival of items from the audiovisual collection and packaging management area (C and D); Tent assembled at the entrance to CDHS to avoid water falling on the items on rainy days (E). *Source* DAD Collection. Photos: Jeferson Mendonça and Vinicius Pequeno

done from August to October 2021. The outcome was zero damage to the collections, and none of the workers caught COVID-19. In all, 190,000 items from the DAD and 80,000 items from the BHCS were moved, involving 68 workers from COC during the process: 11 from the library, 26 from the archives, and 31 from the Museum of Life and Administration sector, besides 20 employees from the Fink moving company. Nonpharmacological measures such as physical distancing, constant hand hygiene, and use of N95 face masks were recommended and used throughout the move, along with immunization of the teams and procedures planned according to the risk management method.

5 Conclusion

The experiences discussed here have demonstrated the importance of implementing risk management as a continuous process that permeates all the preservation activities, contributing to routine preventive conservation measures and replanning in adverse and emergency situations. The results achieved thus far by the Fiocruz teams based on the elaboration of preservation policies and the establishment of a Working Group in preventive conservation and risk management demonstrate the importance of an essential cultural change that allows the definition of strategies focused on prevention, with the safety and security of people and cultural heritage as priorities. The teams' experience has been shared through publications, participation in scientific events, and interaction with students in the executive Master's course and various free courses offered by COC.

An important spinoff of this work was the establishment of a cooperative agreement between Fiocruz, ICCROM, and CCI aimed at promoting progress with innovative tools and educational materials to benefit professional organizations in conservation and restoration. This partnership has resulted in the current development of the ABC Risk Management System for Cultural Heritage, a tool that will support deployment of the ABC Method. We hope that the tool's free availability will contribute to the dissemination of the preventive approach for cultural institutions in various countries.

Considering the impossibility of physical access to the collections, Fiocruz has invested even more in publicizing access to the collections through digital tools such as digitization platforms, open-access digital repositories, digital preservation, interoperable databases, social networks, and web conferences. The current health crisis brought even greater urgency to a previously identified demand for institutions in teaching, research, and cultural memory to adjust to the digital transformations and review their investments in the promotion of solutions for monitoring, automation, and remote control of areas with custody of collections, having preventive conservation and risk management as the central approach.

Acknowledgements The authors wish to thank the members of the Working Group on risk management and preventive conservation at COC and collaborators from other Fiocruz units, especially the Oswaldo Cruz Institute and the Institute of Scientific and Technological Communication and Information in Health. They also thank the Brazilian National Council for Scientific and Technological Development (CNPq) for the science initiation scholarships provided during the pilot cycle of risk management.

References

ABNT—Associação Brasileira de Normas Técnicas. 2018. *ABNT NBR ISO 31000 - Gestão de riscos - Diretrizes.* Rio de Janeiro: ABNT.

Brokerhof, Agnes W. 2006. Canadian museums association cultural property protection. https://www.museums.ca/uploaded/web/docs/CPP.pdf. Accessed 2 December 2021.

Fundação Oswaldo Cruz. 2020. Preservation policy for scientific and cultural collections of Fiocruz. https://www.arca.fiocruz.br/handle/icict/44749. Accessed 4 November 2021.

Fundação Oswaldo Cruz, and Casa de Oswaldo Cruz. 2013. Política de preservação e gestão de acervos culturais das Ciências e da Saúde. https://www.arca.fiocruz.br/handle/icict/15276. Accessed 4 November 2021.

Fundação Oswaldo Cruz, and Casa de Oswaldo Cruz. 2020. Relatório de divulgação dos resultados do primeiro ciclo de aplicação da metodologia de gestão de riscos para o patrimônio cultural da Fiocruz. https://www.arca.fiocruz.br/handle/icict/42316. Accessed 4 November 2021.

Fundação Oswaldo Cruz, and Casa de Oswaldo Cruz. 2021. *Plano de Mitigação dos riscos de transmissão da Covid-19 no ambiente de trabalho da Casa de Oswaldo Cruz.* Rio de Janeiro: Fiocruz/COC.

Instituto Brasileiro de Museus (Ibram). 2020. Recomendações aos museus em tempos de Covid-19. https://www.museus.gov.br/wp-content/uploads/2020/06/Recomendacoes_Museus.pdf. Accessed 2 December 2021.

International Council of Museums—Brasil (ICOM-BR). 2020. Recomendações do ICOM Brasil em relação à Covid-19. http://www.icom.org.br/wp-content/uploads/2020/04/RECOMENDACOES_CONSERVACAO_15_ABRIL_FINAL-1.pdf. Accessed 2 December 2021.

Michalski, Stefan. 2016. Beyond the traditional approach to preventive conservation. Interview to Ruth Bagan. *Rescat* 30: 3–7.

Michalski, Stefan, and Jose Luiz Pedersoli Jr. 2016. The ABC method: A risk management approach to the preservation of cultural heritage. https://www.iccrom.org/sites/default/files/2017-12/risk_manual_2016-eng.pdf. Accessed 4 November 2021.

Risk Management of Archaeo-Palaeontological Collection from Sierra de Atapuerca Sites at CENIEH

Raquel Lorenzo-Cases⬩, Sofía De León-Verdasco⬩, and Pilar Fernández Colón⬩

Abstract The National Research Centre on Human Evolution (CENIEH) hosts a palaeontological collection from Plio-Pleistocene sierra de Atapuerca (Burgos), declared a World Heritage Site by UNESCO in 2000. The aim of the Conservation & Restoration Laboratory of CENIEH is to assure the conservation collection in time, in order to arrest or minimize handling and preserve its high scientific and cultural wholeness. In this manner it becomes a priority to establish specific preventive conservation strategies. Thus, to achieve these goals it is necessary to review and update the CENIEH Plan for Preventive Conservation by developing every stage: documentation, context, risk assessment, proceedings design and verification. Summarizing the efforts of risk management must be focused on making scientific use compatible with conservation criteria.

1 Archaeo-Palaeontological Collection from Sierra de Atapuerca Sites at CENIEH

CENIEH is located in Burgos, in the north of Spain, within a few kilometers from sierra de Atapuerca World Heritage sites. It is part of the Human Evolution Complex, along with Human Evolution Museum, and an Auditorium.

The CENIEH belongs to the Spanish network of Unique Scientific and Technical Infrastructures (ICTS, in Spanish). This refers to large facilities, resources equipment and services necessary for the development of cutting-edge research of the highest

R. Lorenzo-Cases (✉) · S. De León-Verdasco · P. Fernández Colón
CENIEH, Centro Nacional de Investigación sobre la Evolución Humana, Burgos, Spain
e-mail: raquel.lorenzo@cenieh.es

S. De León-Verdasco
e-mail: sofia.deleon@cenieh.es

P. Fernández Colón
e-mail: pilar.fernandez@cenieh.es

Á. F. Perles-Ivars et al. (eds.), *Collection Care*, Springer Proceedings in Archaeology and Heritage, https://doi.org/10.1007/978-3-031-85655-6_19

quality. As a scientific institution its range of activities focus mainly on human evolution during the Late Neogene and Quaternary, which includes collaborative projects at sites and deposits from these periods around the world. In addition, CENIEH is responsible for conservation, restoration, management and recording its archaeological and palaeontological collection. The centre owns forefront technologies, and specialized technical professionals. Laboratories are oriented to different disciplines: Archaeometry, Archaeomagnetism, Geology, Cartography, Bioenergy & Motion analysis, Digital mapping & 3D Analysis, Luminescence dating, Cosmogenic Nuclid dating, Electro Spin Resonance dating, Microscopy & Micro-Computed Tomography, Uranium Series, Prehistoric Technology & Archaeology, and Conservation & Restoration.

One of the exceptional characteristics of the CENIEH is that it hosts heritage from sierra de Atapuerca Plio-Pleistocene sites. Due to this fact, it has a Collections, Conservation & Restoration Area that includes collection management and a Conservation & Restoration Laboratory, something that is usually linked to a museum context. In 2019, the CENIEH signed an agreement with the regional government of Castilla y León to manage Atapuerca collections. This has turned the CENIEH the principal storage center of the palaeontological and archaeological heritage from Atapuerca sites. The agreement also stablishes CENIEH as the institution managing the conservation, restoration and scientific research from these assets. Due to these facts, the number of items safeguarded at CENIEH in the last years has grown from approximately eighteen thousand to more than two hundred thousand. And this number continues growing.

Considering these circumstances, the Collections, Conservation & Restoration Area is facing an exponential increase of its activity, and its needs for material and staff resources. In this way, a refurbishment work took place at the three storage rooms hosted in two floors: An armored room with the most significant scientific collection, and two twin rooms where the main part of the collection is preserved.

Atapuerca sites were established as World Heritage by UNESCO in 2000. Currently, five different species of hominins had been found there. These remains are key specimens in human evolution research and two of them are only found in those sites. Due to this exclusiveness and the high scientific and cultural value, the collection have the highest significance.

The archaeo-palaeontological collection from Atapuerca sites are distributed in three groups: semi-fossil hominin bones such as *Homo antecessor*, *Homo neanderthalensis*, *Homo sp.*, and pre-neanderthals of Sima de los Huesos site; fauna semi-fossil specimens, like *Ursus dolinensis* or *Dolinasorex glyphodon*; and lithic assemblage of various raw materials, like quartzite, silex or sandstone. Palaeontological remains make up a high percentage of the collection.

2 Conservation Needs of the Collection

The main degradation processes of palaeontological heritage cost by from physical properties, chemical composition, and their interaction with soil during burial. Particularly in bones, hygroscopicity and anisotropic nature trigger mechanical weakness in unstable environments. The CENIEH collection general state of conservation is heterogeneous, due to the diversity of archaeological contexts, pre-depositional and post-depositional burial processes, excavation procedures and old restoration treatments. In addition, the main use of collection is scientific research, which implies a real and potential mechanical deterioration risk due to improper handling. Thus, the main conservation needs are related to three factors: nature of collection, scientific use and environmental features at CENIEH. The key guidelines to consider attending to these factors are:

1. Stability of microclimatic conditions, avoiding fluctuations of relative humidity (RH) and temperature (T). Suitable parameter ranges are around: 50% RH and 20 °C, respectively (Michalsky, 2007).
2. Low luminous impact, avoiding IR and UV radiation.
3. Reduce or avoid VOCs (Volatile Organic Compounds) and prevent biological plagues.
4. Adequate physical conditioning that prevents from mechanical damage.
5. Avoid dissociation.
6. Minimize handling.
7. Conservation training for researchers. This is one of the highest challenges: Generally, researchers in human evolution are not aware of the necessity to apply special conservation procedures to the palaeontological heritage.

According to these requirements, the Conservation & Restoration Laboratory prioritizes:

- Accessibility and storage capacity.
- Consultation protocols and upgrading database, to prevent dissociation.
- Care and prevention against potential accidents.
- Maintenance of stable microclimatic conditions.

3 Preventive Conservation and Risk Management of the Collection

It is of utmost importance to find a balance between scientific needs (study and analysis) and heritage conservation. By prioritizing safeguard actions, we avoid losing not only the original material itself but also the scientific information that it could provide in future. In this manner, it is urgent to design specific preventive conservation strategies, starting by risk management.

The aim of the Conservation & Restoration Laboratory is to assure the safe-guarding of the collection in time, preserving its scientific and cultural wholeness for future generations. To manage and preserve a collection of these special features it is necessary to develop risk assessment, which implies identification and evaluation of potential threats based on the magnitude of the most likely risk and its probability (Michalsky, 2004). Taking into account the environmental context at the storage rooms (and their outside influence), a risk classification has been made considering nature and use of the collection. In addition, a classification has been made to asses the cultural significance. Then, we have two scenarios: one for the armored room that is a security room of 50 m^2, and another for the storage rooms 1 and 2, each 120 m^2.

To classify the risk, we followed the guidelines from the *IPCE* (Spanish Cultural Heritage Institute). The two factors that influence the risk assignation are: the probability of happening and the severity of damage. The incidence of these factors is graded on a numerical scale from 1 to 5 that allows to make a quantitative description (IPCE, 2009, IPCE, 2019). The probability of happening could be defined as low, medium, or high. The severity of damage could be light, severe, or very severe, depending on the actions that are needed to stop it: (preventive conservation, remedial conservation or restoration treatments).

Based on the risk evaluation results mentioned before, we included an evaluation of the priority actions. A table shows the priority of each risk from little importance to high importance, depending if the risk does or not need immediate actions (Table 1).

Table 1 Risk evaluation of the armored room

Armored Room—Vault				
Risk	Probability	Severity	Valuation	Priority
Dissociation Physical damage (use & conditioning)	Low Low	Very severe Very severe	3 3	Moderated Moderated
Vandalism	Low	Very severe	3	Not important
Catastrophic events (fire, floods) Incorrect climatic conditioning (RH & T fluctuations, light, pollutants)	Low High	Very severe Very severe	3 5	Moderated Very important
Damage by outside conditions	Low	Light	1	Not important
Biological damage	Low	Light	1	Not important
Negligence in monitoring and control proceedings Inadequate maintenance Absence / Deficiencies of the conservation project, or cultural use	Low Low Low	Severe Light Very severe	2 1 3	Medium Moderated Medium

The evaluation made from the armored room determines that the highest risk is the incorrect environmental conditioning and the consequence of having fluctuating microclimatic conditions: relative humidity and temperature. On the other hand, due to the big size and homogeneity of the collection, the risk assessment made from the storage rooms 1 and 2 determines that dissociation is its highest risk. Below, physical damage by handling and inadequate conditioning is the second concern caused by its scientific use.

The risk assessment provides the path to continue making decisions. To control the microclimatic environment, the access to the storage rooms has been designed through a double door with a middle cabin to stop the current flow between the entrance and the inside room. Furthermore, the air conditioning system is independent from the rest of the building and a device has been installed per storage room. In addition, each air conditioner can be manually adjusted to achieve maximum daily fluctuations over ± 1 °C and $\pm 3\%$ RH. Nevertheless, in the armored room these optimal ranges are not always possible to achieve due to its great influence from external environmental conditions. For that reason, the current target is to adjust the air conditioner to these circumstances and improve its isolation in the near future.

Moreover, a wireless system for data acquisition and monitoring has been installed in the storage rooms. The system sends an alert if detects values of relative humidity and temperature that exceed the optimal range considered for the conservation of collection. Thanks to all these features it is possible to understand the behavior of the storage rooms by making statistical analysis of data.

Consequently, the main goal is to change from an uncontrolled environment influenced by external conditions to a controlled environment (Fig. 1).

Furthermore, risk assessment of the storage rooms 1 and 2, determines that the main objective is to slow down dissociation and physical damage due to scientific studies. With this purpose, we have been designing a physical conditioning with a double purpose: following the conservation criteria and permitting consultation of the collection. The objective is reached by using non-reactive and chemically stable materials, adapting design to each morphology, minimizing handling and avoiding mechanical damage. For this purpose, we have been using storage conditioning materials, known for their suitability:

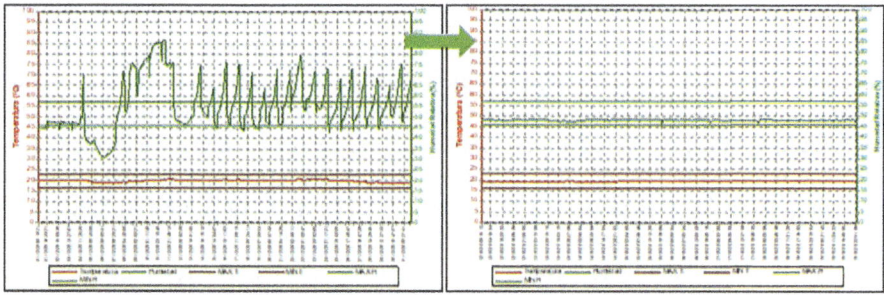

Fig. 1 Comparison of unstable and stable environmental data recorded at the armored room

- Polyethylene foam that absorbs vibrations and shocks.
- *Volara®* and *Tyvek®* in contact with heritage in order to avoiding mechanical injuries like abrasion.
- *Fome-cor®* foam core that provides stability and an aesthetic design.
- Polypropylene stacking boxes with hinged lids and upper side access that protects from dust, radiation and chemical or physical damage.
- Individual polyethylene transparent boxes that provide see-through vision and permits separate specimens.

The design and materials election is the result of experience and conditioning tests made over the years. Clearly, what works in our scientific collection might not work in a different context; for that reason it is important to test conditioning materials in each particular case. A great part of these polymeric materials is submitted to an accelerated aging test thanks to the *POLYVEART* research project from the *IPCE* (Ministerio de Cultura, 2021).

Regarding design and arrangement, we decided to classify palaeontological remains by taxonomy anatomical order. Also, it has been stablished a signal code that avoids risk of dissociation. For small size human remains with a complex morphology, -cranium fragments, vertebra or tooth-, it has been used transparent polyethylene boxes in order to individualize them. This provides visualization and minimize handling. Then, the boxes have been held with *Volara®* to immobilize them inside the storage containers. *Volara®* is a polyethylene foam that is soft, flexible and easy to adapt to the shape needed in each case. Also, for bigger specimens remains it has been created pillow-like holders according to the size of each remain. These holders are made with Tyvek® (high-density spunbound polyethylene fibers) and little polyethylene balls. This solution provides the perfect accommodation for the specific specimen morphology. It is important to create trouble-free solutions that are easy to implement; so the process can be systematized while also taking into account the special characteristics of each fossil (Fig. 2).

Regarding the codification system of storage containers, the content information is located outside the boxes: nature, taxon, raw material, technological description or anatomical element, and topographic reference. The topographic reference provides the location of each container in a storage room, a cabinet, and a shelf. Inside the storage containers lid, a photograph shows the correct content distribution to avoid dissociation.

Conversely, a collection consultation procedure at CENIEH has been defined. This procedure determines loan terms, exhibition and sampling (for destructive and non-destructive analysis). Procedures must systematize loan and consultation demands; they should be a safe channel which singles out requests in order of relevance. The relevance is measured by the importance of the scientific research application and the significance on the requested collection. Besides the consultation demand, a conservation technical regulation mandatory compliance is obligatory: researchers must sign it before having access to the specimens requested. The regulation defines how must be handle the collection and the objects allowed in the consultation room.

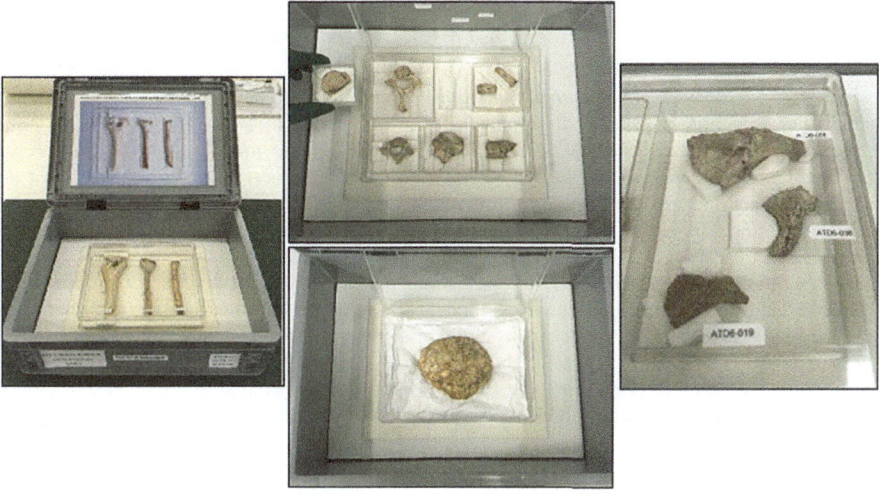

Fig. 2 Physical conditioning of the hominin semi-fossil collection at the armored room

Concerning analysis requests, these are assessed depending on the necessity of taking a sample and if the exam is destructive or non-destructive. Safeguarding of the heritage is always the priority: the Conservation & Restoration Laboratory must approve the analysis by a technical report about the state of conservation, the significance, and the real and potential impact over the preservation in time of the specimen.

4   Conclusions

Risk management is an essential part of the Preventive Conservation Plan. Is the strategy that allows to understand better an institution and learn which are its strengths and weaknesses. At the Conservation & Restoration Laboratory, we are conscious that safeguarding in short, medium, and long terms the Atapuerca world heritage collection is a challenge. Our greatest struggle, beyond achieving an adequate micro-climate, is on risks derived from its large number and scientifical use: dissociation and mechanical damage. Therefore, user training and awareness-raising are our best assets for the future to avoid or minimize the loss of heritage.

We are aware that things that works now could not work in the future; so periodic evaluations of control are a need. Verification is necessary to solve problems and improve the system. Because our global objective is to preserve cultural heritage in time, the best tool to assure we reach the goal is a solid preventive conservation plan.

References

Instituto del Patrimonio Cultural de España, 2019. *Guía para la elaboración e implantación de Planes de Conservación Preventiva*. Madrid: Ministerio de Cultura y Deporte.

Instituto del Patrimonio Cultural de España, 2009. *Normas de Conservación Preventiva para la implantación de sistemas de control de condiciones ambientales en museos, bibliotecas, archivos, monumentos y edificios históricos*. Sección de Conservación Preventiva. Área de Laboratorios. Madrid: Ministerio de Cultura y Deporte.

Michalsky, Stefan. 2004. Care and preservation of collections. In *Running a museum: A practical handbook*, ed. Patrick Boylan, 51–89. París: ICOM & UNESCO.

Michalsky, Stefan. 2007. The ideal climate, risk management, the ASHRAE chapter, proofed fluctuations, and toward a full risk analysis model. In *Experts roundtable on sustainable climate management strategies*, ed. Getty Conservation Institute, 1–19. Tenerife: The Getty Conservation Institute.

Ministerio de Cultura. 2021. Presentación Evaluación de Productos utilizados en Conservación y Restauración de Bienes Culturales Polyevart. Accessed 1st October 2021.

The manufacturer's authorised representative in the EU is Springer
Nature Customer Service Centre GmbH, Europaplatz 3, 69115 Heidelberg,
Germany. If you have any concerns regarding our products, please
contact ProductSafety@springernature.com

Printed and bound by CPI Group (UK) Ltd, Croydon, CR0 4YY
27/04/2026
02097566-0006